Lecture Notes in Mobility

Series editor

Gereon Meyer, Berlin, Germany

More information about this series at http://www.springer.com/series/11573

Tim Schulze · Beate Müller
Gereon Meyer
Editors

Advanced Microsystems for Automotive Applications 2015

Smart Systems for Green and Automated Driving

Springer

Editors
Tim Schulze
VDI/VDE Innovation + Technik GmbH
Berlin
Germany

Gereon Meyer
VDI/VDE Innovation + Technik GmbH
Berlin
Germany

Beate Müller
VDI/VDE Innovation + Technik GmbH
Berlin
Germany

ISSN 2196-5544 ISSN 2196-5552 (electronic)
Lecture Notes in Mobility
ISBN 978-3-319-37018-7 ISBN 978-3-319-20855-8 (eBook)
DOI 10.1007/978-3-319-20855-8

Springer Cham Heidelberg New York Dordrecht London

Softcover reprint of the hardcover 1st edition 2016

Printed on acid-free paper

Springer International Publishing AG Switzerland is part of Springer Science+Business Media
(www.springer.com)

Preface

Green vehicles that provide a high degree of energy efficiency and are powered by electricity or other alternative fuels have been on the agenda of politicians and industry alike since many years now. Public funding of research and innovation in electric vehicles in the European Green Vehicles PPP for example has triggered the development of products, which have been launched successfully to the market recently. At the same time, the topic of automated driving is increasingly raising public attention: highly automated driving, providing yet unsurpassed levels of road safety, efficiency, productivity and social inclusion, seems to be feasible on motorways by the year 2020.

Smart systems combining sensing, cognitive processing and actuation are essential not just for the electrification but also for the automation of the automobile and will remain a subject of research and innovation for quite a while. Particularly for applications in complex environments like city traffic and at higher levels of automation on the way to self-driving capabilities, perception of the driving environment is a challenging task to be addressed, e.g. by multi-sensor systems and sensor data fusion. Furthermore, connectivity and eventually a certain level of artificial intelligence will be needed to ensure the safety of the system. Thus, new links between the smart systems, the Internet of Things and the robotics communities should be created in order to fully embrace the research and innovation needs for the years to come. A first attempt for this has been made earlier in 2015 when the European Technology Platform on Smart Systems Integration (EPoSS) presented a "European Roadmap on Smart Systems for Automated Driving" and contributed to an even broader roadmap activity as a member of the core team of the connectivity and automation task force of the European Road Transport Research Advisory Council (ERTRAC).

The key enabling technologies for the automobile of the future have always been the topic at the International Forum on Advanced Microsystems for Automotive Applications (AMAA) at an early stage. Thus, the topic of the 19th AMAA 2015, held in Berlin on 7–8 July 2015, is "Smart Systems for Green and Automated Driving". The AMAA organisers, VDI/VDE Innovation + Technik GmbH together with EPoSS, greatly acknowledge the support given for this conference, particularly from the European Union through the Coordination Action "Global Opportunities for Small and Medium Sized Enterprises in Electric Mobility" (GO4SEM).

The papers in this book, a volume of the Lecture Notes in Mobility book series by Springer, were written by leading engineers and researchers who have attended the AMAA 2015 conference to report their recent progress in research and innovation. The papers were peer-reviewed by the members of the AMAA Steering Committee and are made accessible worldwide. As the organisers and the chairman of the AMAA 2015, we would like to express our great appreciation to all the authors for their high-quality contributions to the conference and also to this book. We would also like to gratefully acknowledge the tremendous support we have received from our colleagues at VDI/VDE-IT.

Berlin
June 2015

Tim Schulze
Beate Müller
Gereon Meyer

Advanced Microsystems for Automotive Applications 2015

Funding Authority

European Commission

Supporting Organisations

European Council for Automotive R&D (EUCAR)
European Association of Automotive Suppliers (CLEPA)
Strategy Board on Electric Mobility (eNOVA)
Advanced Driver Assistance Systems in Europe (ADASE)
Zentralverband Elektrotechnik- und Elektronikindustrie e.V. (ZVEI)
Mikrosystemtechnik Baden-Württemberg e.V.

Organisers

European Technology Platform on Smart Systems Integration (EPoSS)
VDI/VDE Innovation + Technik GmbH

Steering Committee

Mike Babala, TRW Automotive, Livonia, MI, USA
Serge Boverie, Continental AG, Toulouse, France
Geoff Callow, Technical & Engineering Consulting, London, UK
Kay Fürstenberg, Sick AG, Hamburg, Germany
Wolfgang Gessner, VDI/VDE-IT, Berlin, Germany

Conference Chair

Contents

Part I
Driver Assistance and Vehicle Automation

Autonomous Parking Using Previous Paths

Christoph Siedentop, Viktor Laukart, Boris Krastev, Dietmar Kasper, Andreas Wedel, Gabi Breuel and Cyrill Stachniss

Abstract This paper is about mapping the drivable area of a parking lot for autonomous parking. Manual map creation for automated parking is often impossible, especially when parking on private grounds. One aspect is that the number of private properties is very large and private parking should not be included in public maps. The other aspect is that an owner and operator of a car often has very specific ideas of where the car may be driven. Our approach creates maps using just the previously driven paths. We describe the drivable area through triangles using established methods from Computer Graphics. These triangles are generated by overlaying circles of a certain radius over the driven paths. These circles create a so-called alpha-shape and approximate the drivable area. The description through

---

V. Laukart developed the methods in relation to α-shapes and the Delaunay triangulation. B. Krastev enabled the search. C. Siedentop wrote the paper and provided the evaluation.

C. Siedentop (✉) · V. Laukart · B. Krastev · D. Kasper · A. Wedel · G. Breuel
Daimler AG, Autonomous Driving, HPC G023, 71059 Sindelfingen, Germany
e-mail: Christoph.Siedentop@daimler.com

V. Laukart
e-mail: Viktor.Laukart@daimler.com

B. Krastev
e-mail: Boris.Krastev@daimler.com

D. Kasper
e-mail: Dietmar.Kasper@daimler.com

A. Wedel
e-mail: Andreas.Wedel@daimler.com

G. Breuel
e-mail: Gabi.Breuel@daimler.com

C. Stachniss
Institute for Geodesy and Geoinformation, University of Bonn, Nussallee 15, 53115 Bonn, Germany
e-mail: cyrill.stachniss@igg.uni-bonn.de

T. Schulze et al. (eds.), *Advanced Microsystems for Automotive Applications 2015*, Lecture Notes in Mobility, DOI 10.1007/978-3-319-20855-8_1

triangles ("Delaunay triangulation") allows for fast retrieval and easy expansion with new paths. Finally, a simple conversion of the triangulation into a Voronoi diagram enables fast path searching. In this paper we thus present an efficient framework for determining drivable areas and allows searching for a drivable path. Finally, we show that this method enables real-time implementation in an autonomous car and can cope with new obstacles at planning time.

Keywords Autonomous parking · Mapping · Drivable area · Delaunay triangulation · Alpha shapes

1 Introduction

Each owner and user of private property will have individual and varying notions on where she would like her car to drive. Especially for parking it may be perfectly acceptable for one user to drive across grass, and not desirable at all to another driver. This example illustrates how there cannot be one set of rules determining drivable areas. We solve this problem by learning from the driver where the car is allowed to drive. On top of that our approach creates a safeguard on all other sensors that the vehicle might have.

As we hope the benefits are obvious, the challenges should not be taken lightly. We wish to traverse the area as few times as possible but need to integrate new traversals into the drivable area. Finally, our solution must be suitable for a real-time system.

The task can be summarised in the following four steps:

1. Taking a set of reference driving through the terrain and noting where the car was parked.
2. Create a map with the drivable terrain.
3. Plan a new path from a start position to one of the parking spots.
4. Avoid obstacles and re-plan the path as obstacles appear.

In the next section we describe existing research for the task of determining drivable areas. Section 3 describes our approach to this task and makes up the primary contribution of this paper. In Sect. 4 we describe our experimental platform and the experiments we undertook. This is followed by a results and an outlook section.

2 Preliminaries

2.1 Explanation of Terms

2.1.1 Duality in Graphs

Duality in graphs mean that one graph can be converted into the other. The general concept is more refined but for this paper the following explanation may be used: The graph of the Delaunay triangulation becomes the Voronoi graph by placing a vertex inside each triangle and connecting vertices of neighboring triangles with an edge.

2.1.2 A* Search

The A* search algorithm is the graph search algorithm when a heuristic exists. It improves the well-known Dijkstra's algorithm [3, Chap. 24.3] by the use of a heuristic. A heuristic is an oracle that can guess the cost to the goal. A valid and often used heuristic is the straight-line distance to the goal. The A* algorithm is complete in that it will always find a solution if one exists and optimal in that it will find the shortest path in the fastest way possible (measured by nodes expanded and without preprocessing). Necessary is only an admissible heuristic. For an exemplary overview the reader is referred to Wikipedia (*A* search algorithm*) or [15, Sect. 3.5.2] of "Artifical Intelligence - a modern approach". The original paper is [10].

2.2 Related Research

Classifying drivable and non-drivable terrain has often been the focus of robotics research: in the context of all-terrain robots [13, 17], and even autonomous driving [4]. What all these approaches have in common is their effort to predict drivability. This is either done through camera based systems, LIDAR, IR-distance sensors or a combination thereof. We, however, are interested in sensor-less drivability classification. First, because none of these approaches can guarantee perfect precision. For example, Poppinga et al. [13] achieve at best 100 % accuracy but at worst an unacceptable 83 %.

Secondly, we want our approach to work independently of LIDAR. As such the only other ground-sweeping sensors are cameras. Radar and ultrasonic sensors ignore the ground unless mounted in different configurations.

Our usage of α-shapes to compute a drivable map is motivated by several papers in the field, which show their geometric and computational desirability. McCarthy et al. use α-shapes as a map to determine when a path does not exist [11]. Alzantot and Youssrf [2] describe indoor floor plans using—among other things—α-shapes.

A treatment on necessary sampling conditions (specifically density) of points to construct α-shapes is given by Sakkalis and Charitos in [16].

Worral et al. [18, Fig. 6] use the heading information to generate road maps using α-shapes. They adapt the circumcircle test such that points with a heading difference above 90 degrees are ignored. With their method different lanes are constructed which can be used to determine intersections.

2.3 System Setup

The vehicle is equipped with localising sensors. From a combination of DGPS, Radars and cameras a vehicle position is found to an accuracy of 2 cm. Our approach, however, does not depend on any specific accuracy. Instead, any accuracy in localisation propagates through our algorithm.

In addition to the presented algorithm, there is also working obstacle detection. This is a security layer that informs the car of non-static objects that would be encountered from day to day. We later show that new points can be added at runtime easily. Obstacles detected by the sensors are added at runtime and planned around in real time (Fig. 5c).

The vehicle is operating in a publicly accessible parking area, such as a corporate car park, or a single parking spot on domestic property. We tested our approach in both situations.

3 Method

With localisation we collect the vehicle's travelled path which we describe through a left and right polyline. This can be sampled to generate a set of points.

A good representation for these points, which is often employed in Computer Graphics, is to describe the shape as a triangulation. The Delaunay triangulation [6] has good and stable properties, so it is often used for this purpose. These good properties mean that it avoids "skinny" triangles and is the dual to the Voronoi Diagram.

In our approach we first determine a Delaunay triangulation (Sect. 3.1). The boundaries of the drivable area is computed using an α-shape algorithm (Sect. 3.2). Once the Delaunay triangulation (DT) has been cleaned up, the dual to the DT is a Voronoi graph that gives us a searchable graph for path planning (Sec. 3.3). Instead of searching through the state space for the vehicle, we search in its 2D-projection into an Euclidean map—the above Voronoi diagram. We retrieve several paths and check for the optimal trajectory among them. We do so to facilitate various optimality criteria, like maximum distance to objects, minimized jerk, and others. The details of this can be seen in Sect. 3.4.

3.1 From Trajectory to Delaunay Triangulation

Having driven the vehicle around the parking lot or in the vicinity of a single parking space, the localisation module provides us with a number of trajectories. Each trajectory is a set of tuples describing the locations of the vehicle x_i, y_i, its headings θ_i and the corresponding time t_i. It is assumed that the area the car covered while recording the trajectory is safe for driving.

To determine a drivable region these trajectories have to be converted into a representation of the drivable area. Grid approaches are conceivable but we decided to describe the area through polygons. This makes it scale independent with low memory consumption, and in our case allows fast querying.

The polygonal description was picked to be α-shapes [2].

In the following we describe the algorithm employed to generate a so called Delaunay triangulation [6]. The DT is a suitable input to the α-shape algorithm.

Given a set of driven trajectories $\mathcal{T}$ with each trajectory consisting of configuration tuples (x, y, θ, t), we generate a set of points $\mathcal{P}$ relating to the inscribed rectangle of the car. A given configuration is thus transformed to four points.

Next the Delaunay triangulation $DT(\mathcal{P})$ is found on these points $\mathcal{P}$. For this there exists a choice between four algorithms. *Randomised incremental approach*, *Divide & Conquer*, *Sweep Line*, *Sweep hull*, and the *Flip* algorithms [12]. The *flip algorithm* is quadratic in complexity, and is here ignored. Furthermore, the divide and conquer and the sweepline algorithm suffer from numeric instability. We chose the randomised incremental approach [5] because it offers the same $O(|\mathcal{P}| \log |\mathcal{P}|)$ complexity as most other algorithms but additionally allows cheap insertion of further points. This is a property that will be beneficial later on.

Before the algorithm is started, duplicate points must be removed. We do this by lexicographically sorting points and then removing duplicates.

The defining feature of the Delaunay triangulation is that for any two adjacent triangles a flip of the shared edge will not result in a larger minimal angle. Compare Fig. 1 for an example.

3.2 From Delaunay-Triangulation to α-Shapes

The Delaunay triangulation is only an intermediary step to determine the drivable area. α-Shapes are best understood through a special case, namely the convex hull. This case occurs as $\alpha \to \infty$. The more general α-shapes allow concave segments to the hull and even topological holes. The parameter α controls how indented the concave indents may become.

Formally correct, if and only if a disc with radius α can be placed in the plane such that no points lie within it, that dics is not part of the α-hull [8]. Points on the α-hull define the vertices of the α-shape.

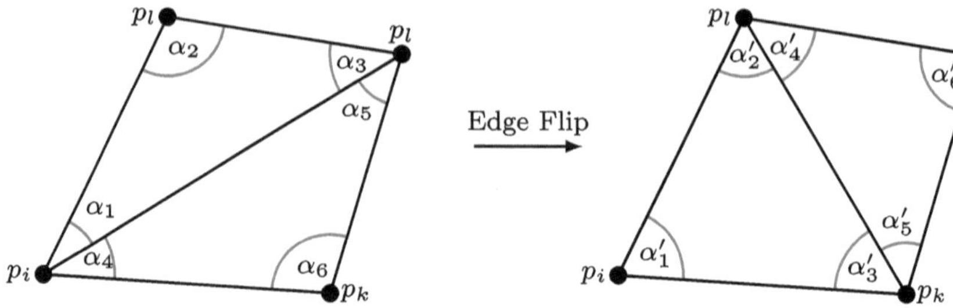

Fig. 1 The Delaunay criterion requires that two adjacent triangles *maximise* the *minimal interior angle*. A simple flip of the shared edge restores the criterion. Specifically, in the *left figure* α_1 is the smallest angle. In the *right-hand figure* α'_4 has become the smallest angle. Because $\alpha'_4 > \alpha_1$ the flip of the diagonal from the *left to the right figure* maximised the minimal angle. Therefore the *right-hand side* is the correct triangulation

The requirements for α-shapes are (1) sufficiently dense sampling of points, and (2) reasonably uniform distribution of points. Both of these requirements are easily met in our application by sampling in uniform distances along the trajectories.

Given a Delaunay triangulation the α-shape can be constructed in a straightforward manner. For each triangle in DT the circumcircle is checked for its radius. If the radius is larger than α, the triangle is not part of the α-shape. Finally, all outer points and its edges are found by marking all triangles as unvisited and then visiting all triangles and recursively marking all neighbours as visited. Only outer vertices need to be kept. Because every triangle is visited exactly once the complete α-shape algorithm possess $O(|\mathcal{P}|)$ complexity.

3.3 Delaunay as the Dual to Voronoi: Easy Search

Some readers will know that the Delaunay triangulation (DT) is the dual to the Voronoi graph. After having purged undrivable areas from the DT, all the triangles make up the drivable area. As such the Voronoi encompasses all, but not more, of the drivable area.

The path along the vertices of the Voronoi diagram can be converted to a corridor as constraints for the optimisation: Selecting the dual edges of the Delaunay triangulation provides us with the left and right boundaries of the corridor.

3.4 K*-Shortest Paths

The A* algorithm (see Sect. 2.1) provides the optimal solution given a suitable modeling of the query. For computational reasons we found that searching in the

2D-space through the Voronoi graph gives a solution faster and within the embedded environment of an autonomous street vehicle. It is self-evident that this will not guarantee an optimal solution. By evaluating several local optima of the 2D-search graph we believe that one of them will provide the global optimimum after optimisation. In practise this yields subjectively optimal paths.

This approach is also suitable when not all costs can be integrated into the search graph. In our case these are, among others, dynamic costs such as lateral and longitudinal acceleration and jerk.

Finally, providing multiple possible paths potentially allows an occupant to make a subjective and individual selection of the path to be taken.

We implemented this through a combination of the A* algorithm [10] with the K-shortest loopless path algorithm from Yen [19]. K-shortest loopless paths was extended by Eppstein [9] and both algorithms were combined by Aljazzar and Leue [1] to K^* (Fig. 2).

In Fig. 3 the multiple different solutions are shown.

The multiple solution can then be checked for the final orientation of the car. Those found suitable can be optimised with a local, convex cost function [20]. Of those, the lowest final cost is used and passed to the control system.

4 Evaluation and Results

The presented approach was tested on a Mercedes-Benz E-Class fitted with radar sensors and a stereo camera system. A similar system is described in Dickmann et al. [7]. For this paper it suffices to know that there is complete coverage from radar sensors. We can also build on functioning localisation and obstacle detection.

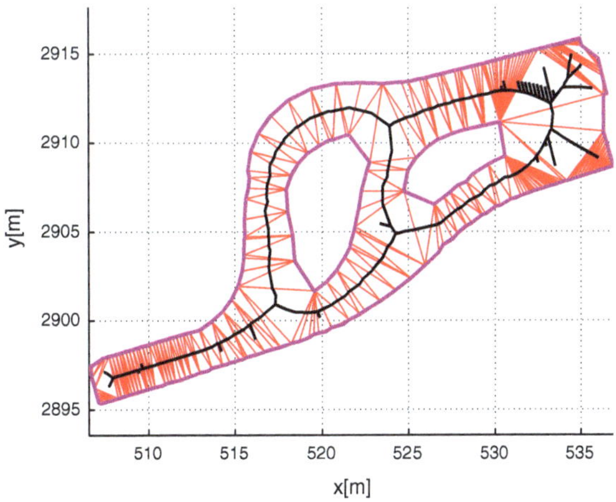

Fig. 2 A parking scenario: The computed drivable area is shown with a *magenta* outline. The Delaunay Triangulation after α-shape cleanup is shown in *red*. Finally, the *black* lines visualise the Voronoi Diagram

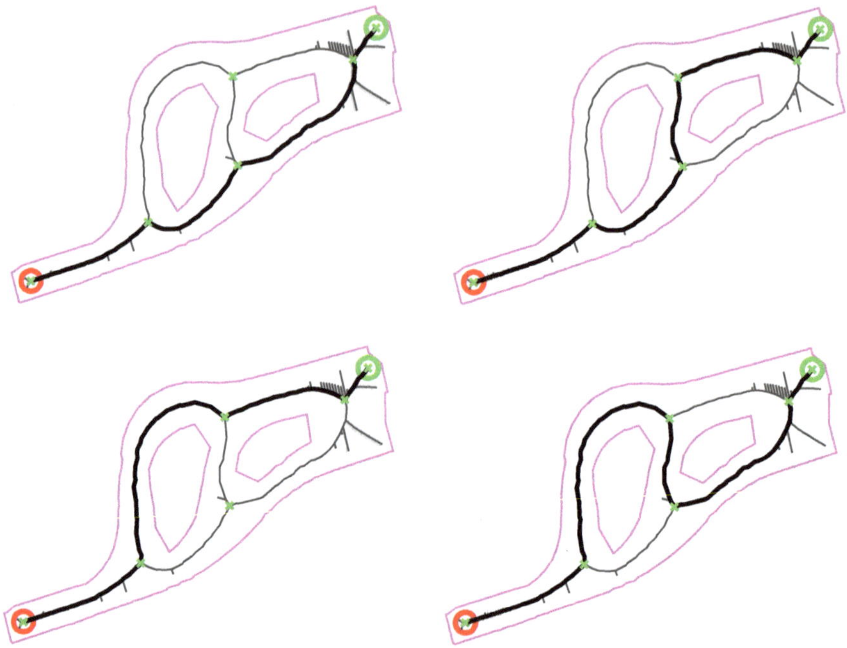

Fig. 3 An example of *k shortest paths* returned for a parking scenario (compare Fig. 2). The searched path is from the *large red circle* to the *large green circle*. *Small green crosses* represent a road graph of the parking lot. The *black lines* indicate the four shortest path along the Voronoi Diagram (shortest *top left*, longest *bottom right*). The *thin magenta line* represents the drivable area boundaries and the dimension is approximately 35 m by 25 m

Our measurements produce good results on two different parking lots of around 50 m × 100 m in size. Please refer to Figs. 4 and 5.

Because of its real-time capabilities the algorithm can cope with obstacles as they appear. When that happens, the α-shape is updated and a new corridor is calculated through the Voronoi diagram. All of this happens at run-time.

As our approach assumes that traversed terrain is safe for driving on, there are cases where the assumption and thus the algorithm could fail. Imagine a structure as typically found in car repair shops. Two narrow planks that the car drives over to be inspected from below. Similar structures can occasionally be found in parking lifts. Here a lateral offset of a few centimeters would turn the same trajectory undrivable. It is important to know that our system cannot cope with such a situation. In most circumstances, however, a parking vehicle will drive where the ground between the left and right tires is drivable as well.

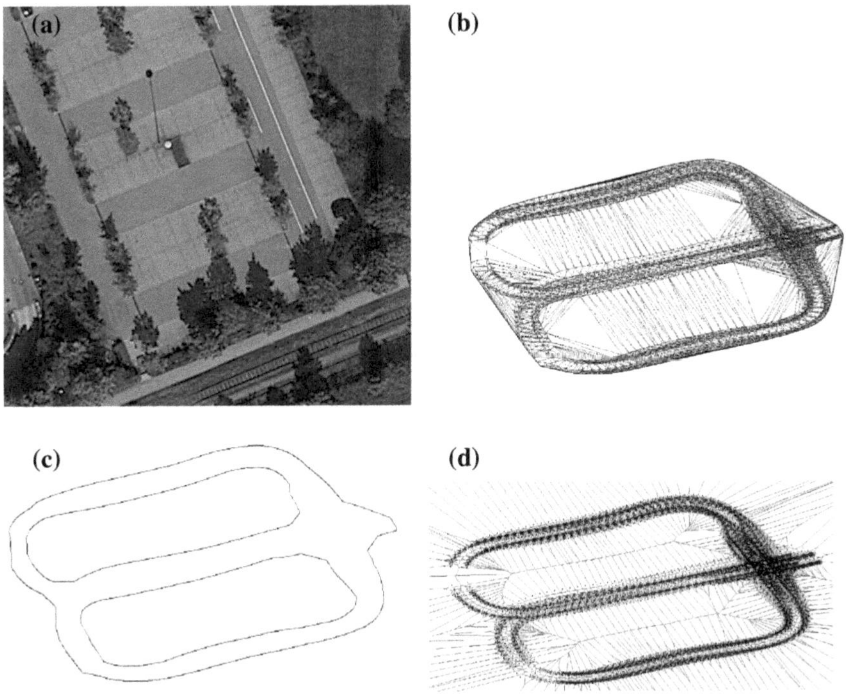

Fig. 4 The steps from driving the trajectory to the final Voronoi diagram. **a** An aerial view of the parking lot (© Google Maps 2015). **b** Delaunay triangulation. **c** α-shape of the same parking lot with an α value of 2.08 m exactly the vehicle width. **d** Voronoi diagram of the parking lot

5 Outlook

There are two omissions in the approach. The first is that the approach cannot detect very thin static obstacles. We call this the thin-fence-problem. If two trajectories were driven parallel to a fence on both sides of the fence the algorithm would falsely classify the railing as drivable.[1]

The second omission is when used as a tool for search. Here our solution is necessary for search but not sufficient. For example, for forward-backward maneuvering we rely solely on the local planner and no costs are taken into consideration during search through the Voronoi. In more general terms, the search through the Voronoi graph generates path candidates. It does not, however, search through the state space of the vehicle which would need to include at least orientation and driving directions.

[1]Points will be generated at $\alpha/2$ to each side of the two center lines. If the paths are less than 2α apart the inner points will be less than α apart and hence merged in the α-shape.

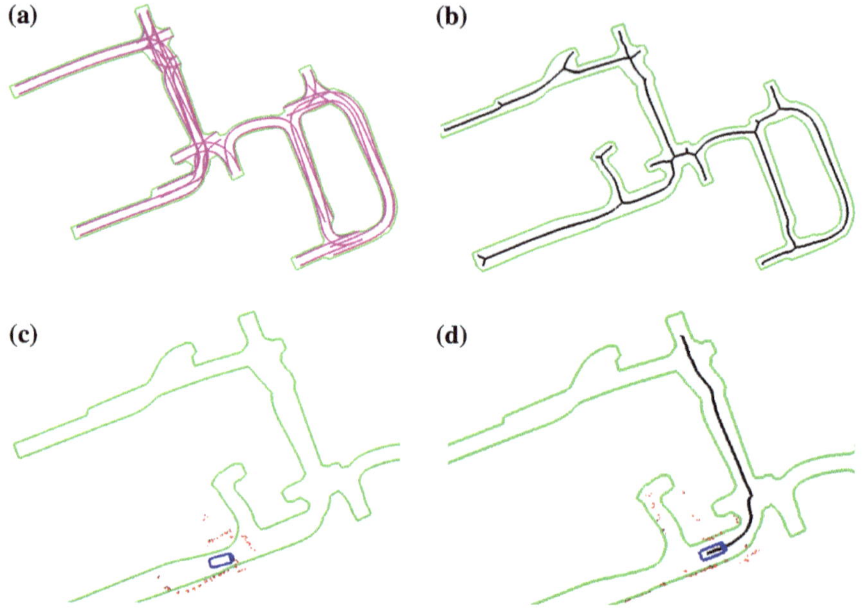

Fig. 5 A different example. **a** Drivable area creation through driven trajectories. **b** *Black center line* shows the Voronoi Diagram. **c** Drivable area has some new obstacles shown here as *red dots*. **d** The vehicle has selected a traversable path

On the other hand, there are many further developments that would benefit from our solution. For example, Ziegler et al. [21] use a Voronoi diagram as a heuristic for the obstacle-aware distance through the map and amend it with a vehicle-kinematic-aware obstacle free distance function as heuristics. It may be beneficial to use the structure of the Voronoi to add an adaptive grid. Rufli and Siegwart [14] have suggested using the major axis of an unstructured environment. We would suggest using the Voronoi diagram to provide a more informed structure than the major axis.

6 Conclusion

In this paper we developed an algorithm which enables a car to autonomously park itself without the prior existence of a map. Our focus was on determining the drivable area and planning a suitable path. We presented an application of the Delaunay triangulation in combination with α-shapes that generates a suitable map representation. We believe that our novel mapping approach is not only helpful in this parking situation but also at other low speed scenarios, especially those where we can rely on sensors to detect drivability in previously visited areas. The benefit of the presented solution is to guarantee never to drive where a user has not been before.

By relying only on driven trajectories and combining this "sensor" with the α-shape algorithm we have shown a feasible system that enables a car to park itself autonomously. Furthermore, our approach naturally lends itself as a structure for path planning and future research: both global planning over the resulting Voronoi diagram becomes simple, and local trajectory optimisation is provided with the appropriate constraints.

We have implemented our approach in an autonomous vehicle. Experimental results show that non-expert drivers are able to teach the vehicle where to drive and where to park.

References

1. Aljazzar H, Leue S (2011) K*: a heuristic search algorithm for finding the k shortest paths. Artif Intell 175(18):2129–2154
2. Alzantot M, Youssef M (2012) Crowdinside: automatic construction of indoor floorplans. In: Proceedings of the 20th international conference on advances in geographic information systems. ACM, pp 99–108
3. Cormen TH, Leiserson CE, Rivest RL, Stein C (2009) Introduction to algorithms, 3rd edn. The MIT Press, Cambridge
4. Dahlkamp H, Kaehler A, Stavens D, Thrun S, Bradski GR (2006) Self-supervised monocular road detection in desert terrain. In: Robotics: science and systems. Philadelphia
5. de Berg M, Cheong O, van Kreveld M, Overmars M (2008) Computational geometry: algorithms and applications. Springer Science & Business Media, Berlin
6. Delaunay B (1934) Sur la sphere vide. Izv. Akad. Nauk SSSR, Otdelenie Matematicheskii i Estestvennyka Nauk 7(793–800):1–2
7. Dickmann J, Appenrodt N, Bloecher H-L, Brenk C, Hackbarth T, Hahn M, Klappstein J, Muntzinger M, Sailer A (2014) Radar contribution to highly automated driving. In: 11th European radar conference (EuRAD). IEEE, pp 412–415
8. Edelsbrunner H, Kirkpatrick D, Seidel R (1983) On the shape of a set of points in the plane. IEEE Trans Inf Theory 29(4):551–559
9. Eppstein D (1998) Finding the k shortest paths. SIAM J Comput 28(2):652–673
10. Hart PE, Nilsson NJ, Raphael B (1968) A formal basis for the heuristic determination of minimum cost paths. IEEE Trans Syst Sci Cybern 4(2):100–107
11. McCarthy Z, Bretl T, Hutchinson S (2012) Proving path non-existence using sampling and alpha shapes. In: IEEE international conference on robotics and automation (ICRA). IEEE, pp 2563–2569
12. Mehlhorn K, Näher S (1999) LEDA: a platform for combinatorial and geometric computing. Cambridge University Press, Cambridge
13. Poppinga J, Birk A, Pathak K (2008) Hough based terrain classification for realtime detection of drivable ground. J Field Robot 25(1–2):67–88
14. Rufli M, Siegwart R (2010) On the design of deformable input-/state-lattice graphs. In: IEEE international conference on robotics and automation (ICRA). IEEE, pp 3071–3077
15. Russell S, Norvig P (2009) Artificial intelligence: a modern approach, 3rd edn. Prentice Hall Press, Upper Saddle River
16. Sakkalis T, Charitos Ch (1999) Approximating curves via alpha shapes. Graphical models and image processing 61(3):165–176

17. Schadler M, Stückler J, Behnke S (2014) Rough terrain 3d mapping and navigation using a continuously rotating 2d laser scanner. KI-Künstliche Intelligenz 28(2):93–99
18. Worrall S, Orchansky D, Masson F, Nieto J, Nebot E (2010) Determining high safety risk scenarios by applying context information
19. Yen JY (1970) An algorithm for finding shortest routes from all source nodes to a given destination in general networks
20. Ziegler J, Bender P, Dang T, Stiller C (2014) Trajectory planning for berthaa local, continuous method. In: Intelligent vehicles symposium proceedings. IEEE, pp 450–457
21. Ziegler J, Werling M, Schroder J (2008) Navigating car-like robots in unstructured environments using an obstacle sensitive cost function. In: Intelligent vehicles symposium. IEEE, pp 787–791

Dynamic eHorizon with Traffic Light Information for Efficient Urban Traffic

Hongjun Pu

Abstract The electronic horizon (eHorizon) is an emerging technology supporting advanced driver assistance systems (ADAS) with respect to fuel efficiency and road safety. Using the static road attributes provided by the eHorizon, new kinds of ADAS applications become possible. Considering the factor that the traffic dynamics, especially in intersection areas, has also a significant impact to the fuel consumption and traffic efficiency, this paper is devoted to the extension of the current products of eHorizon with traffic light information. It introduces a concept for the presentation of the traffic light information in accordance to the ADASIS v2 specification, so that today's subscribers of eHorizon, i.e. the ECUs with ADAS applications, can use the dynamic data with only minimal modification.

Keywords ADAS · eHorizon · Traffic light · Energy efficiency · Traffic efficiency · Road safety

1 Introduction

For diverse reasons, road traffic will remain one of the greatest challenges in the coming years, especially in and around cities. This is because of the globalization of the world economy, the urbanization of developing countries and the re-urbanization of industrial countries due to aging population. Also the individualization of the consume markets, e.g. e-commerce, causes additional traffic related to ware-delivery. All these factors lead to increased road traffic and requirements on intelligent solutions for traffic and energy efficiency.

H. Pu (✉)
Continental Automotive GmbH, Philipsstrasse 1, 35576 Wetzlar, Germany
e-mail: hongjun.pu@continental-corporation.com

T. Schulze et al. (eds.), *Advanced Microsystems for Automotive Applications 2015*,
Lecture Notes in Mobility, DOI 10.1007/978-3-319-20855-8_2

The electronic horizon (eHorizon) is an emerging technology that supports advanced driver assistance systems (ADAS) with respect to fuel efficiency and road safety. Current products of eHorizon extract information from a geo-database and provide certain road attributes like intersection, slope, curvature, speed limit and lane information over a well specified CAN-interface.

In this way, nearly all ECUs in the vehicle can get aware of the roads ahead, as if they have an electronic "eye" observing the horizon. Using the data of eHorizon, new kinds of driver assistance functions for energy-efficiency and road safety become possible. For example, a predictive vehicle cruise control of the truck manufacturer Scania using road geometry information can save up-to 3 % fuel consumption compared to a convenient cruise control [1]. In similar ways, the hybrid and electrical vehicles are expected to achieve more energy-efficiency by deploying a driving and charging strategy adaptive to road properties.

Although predictive road information contributes a lot to the optimization of the operations of gear box, fuel supply and braking system, no longitudinal control for urban traffic would be green enough, as long as the traffic dynamics in intersections are not involved. The vehicle motion at intersections, governed by the traffic light, has a significant impact to the total fuel consumption (or energy balance for electrical and hybrid vehicles) of a city route.

Considering the above intersection problem, this paper presents a solution to provide traffic light information in intersection areas by extension of the eHorizon. After a short over view of eHorizon we will discuss the technologies enabling the transmission of traffic light information into the vehicles. Then, as main contribution of the paper, a concept for presenting the traffic light status and phases on the eHorizon will be introduced and discussed. We will illustrate how the dynamic traffic light information can be embedded in the standard structure of eHorizon, so that the subscribers of eHorizon data, i.e. the electronic control units (ECU) with ADAS applications, need only minimal modification in order to use this dynamic information.

2 The eHorizon

The development of eHorizon was initialized by the ADASIS Forum established in 2001 by a group of car manufacturers, in-vehicle system developers and map data companies. Goal of the ADASIS Forum was to develop a standardized interface between digital map and ADAS applications. The first system and interface specification of ADASIS Forum, e.g. ADASIS v1, was worked out within the EU-funded project PReVENT/MAPS&ADAS. The current version, e.g. ADASIS v2 [2], are widely accepted by the automotive industry as de facto standard. Ress et al. published in 2008 a well structured description of ADASIS v2 [3].

2.1 Stubs and Paths

Paths and stubs are the basic elements of eHorizon describing the road segments and the branch-points. As shown in Figs. 1 and 2, a path with a path-ID represents a road link and stubs are defined at the intersections and branching-points on the path. Sub-paths with its own path-ID may start from each stub.

As the name implicates, only road links on the horizon, i.e. ahead of the ego-vehicle and are reachable in reasonable time, are presented as paths of the eHorizon. Further, the paths are one-dimensional, i.e. all road attributes are positioned by their relative distances to the origin of the path.

2.2 ADAS Attributes

While the paths and stubs present the basic geometry of the road, the useful information for the ADAS applications are provided by the so called ADAS attributes. These are the road classes, scopes, curvatures, speed limits, etc. defined on certain positions and segments of a path.

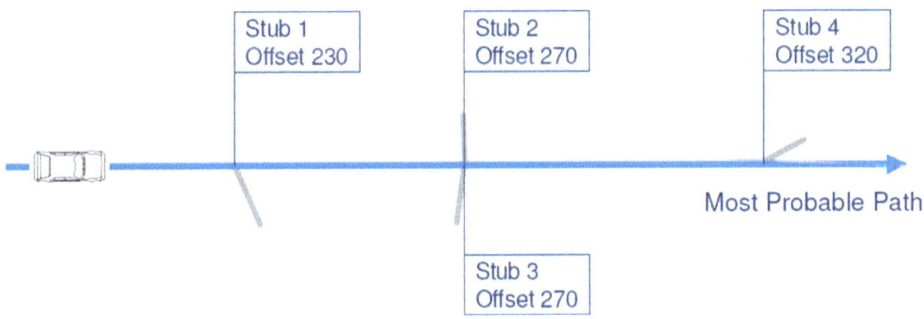

Fig. 1 Stubs of the eHorizon [3]

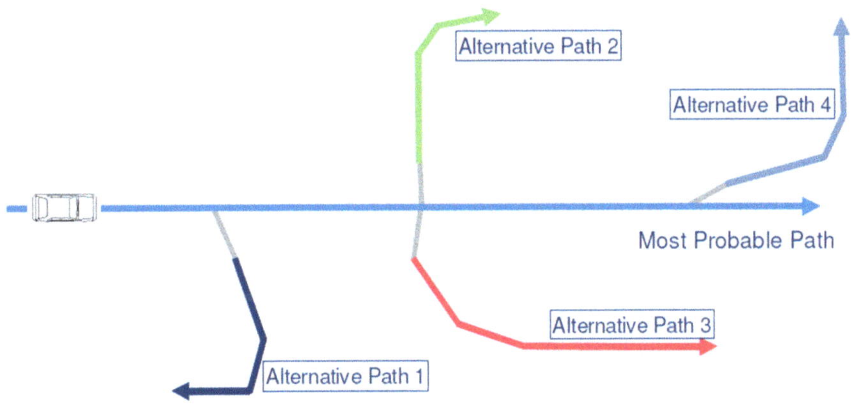

Fig. 2 Sub-paths starting from stubs of the eHorizon [3]

2.3 The Most Probable Path (MPP)

Mainly based on the road classes, possibly also determined by other criteria e.g. navigation guidance, the eHorizon is able to provide a so called most probable path (MPP) indicating the most likely route of the ego-vehicle in near future. The stubs and sub-paths are preferably created along with the MPP. The MPP is continuously checked against the vehicle position and re-created as soon as the ego-vehicle is leaving the current MPP.

2.4 Access to eHorizon via CAN Bus

The ADAIS v2 protocol defines the way to access to the eHorizon via CAN. The choice of CAN was well reasoned from the point of view of the practice, since CAN is the most widely accessible bus system in vehicle.

All eHorizon data are presented in 64-bit CAN frames. Due to the factor that CAN-bus possesses just a limited band-width, the use of CAN messages must be very efficient, e.g. to provide as much as possible information with as less as possible messages.

3 Dynamic Information on eHorizon

Current eHorizon products provide to the ADAS applications static road information extracted from a geo-database. Because of the great significance of dynamic traffic information to the traffic and energy efficiency, there were a serial of publications [4–6] in the past years on the integration of dynamic contents in eHorizon. The presented solutions, however, are of generic nature and did not include concrete ways and steps for the realization.

This paper, in comparison to the previous ones, presents a practical concept to bring traffic light information into the eHorizon in accordance with ADASI v2. Before the solution will be introduced in the following Sect. 4, we firstly discuss in this section the technologies of vehicles communication that make the dynamic traffic light information available in the vehicle.

3.1 Traffic Light Information via ITS-G5 (C2X)

Via C2X communication (ITS-G5/IEEE802.11p), the traffic light control unit can directly send a message named “Signal Phase and Timing” (SPAT) to the vehicles in a range of some hundreds of meters.

Usually, a SPAT message is dedicated for a traffic light governing certain road lanes in an intersection and contains information of the status and the next changes of this traffic light. The geometric information of the intersection, e.g. ingresses, egresses and lanes are described by another C2X message called “INTERSECTION”. Each SPAT message has to be related to an INTERSECTION message. And an INTERSECTION message may be referenced by many SPAT messages.

Since the standardization of the C2X messages “INTERSECTION” and “SPAT” is still ongoing, the exact contents of the messages are subjected to further changes. However, the basic information depicted in the following Fig. 3 can be expected from a SPAT message.

3.2 *Traffic Light Information as TPEG-TSI*

Traffic light information can also be presented in a dedicated TPEG-message, e.g. Traffic Signal Information (TSI). Like other TPEG-messages, the TSI may be transmitted via different channels, e.g. Internet, DAB, etc.

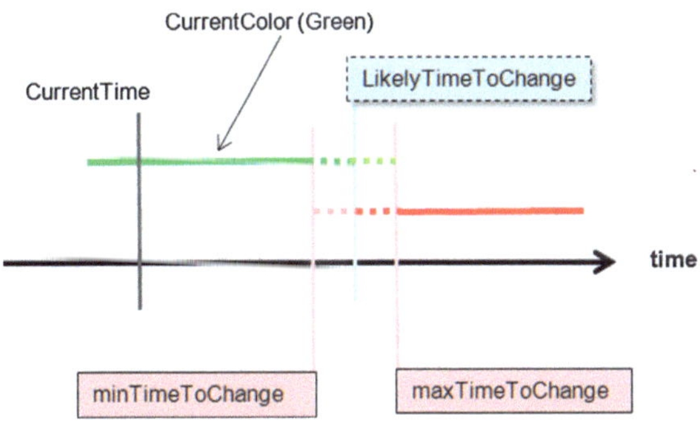

Fig. 3 Traffic light information involved in a C2X SPAT message

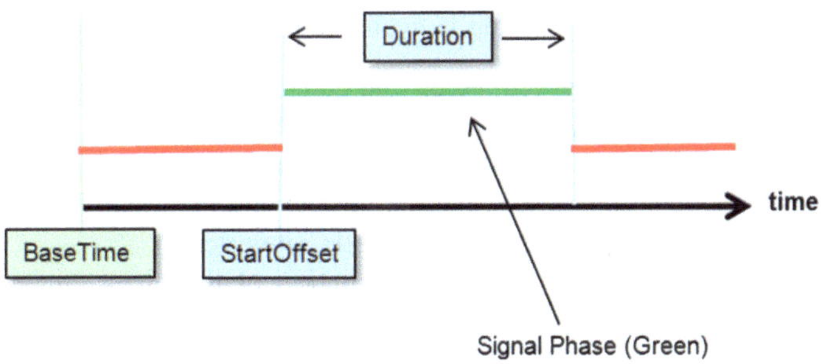

Fig. 4 Traffic light phase information of fixed time control

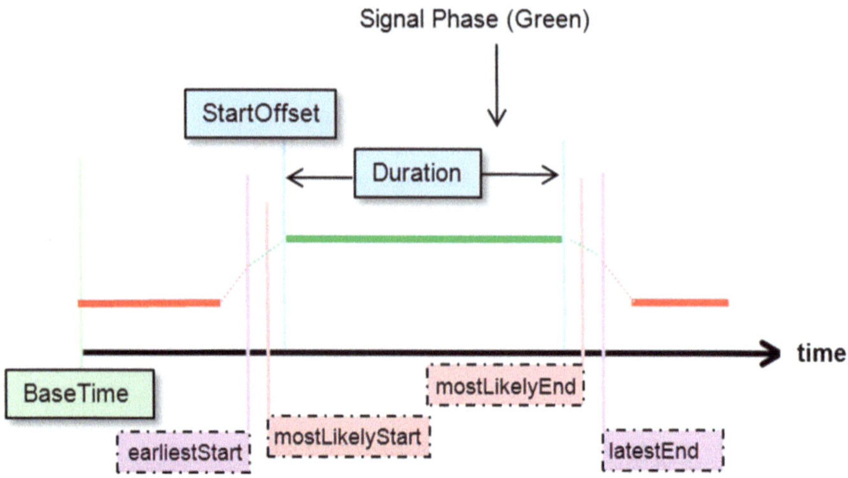

Fig. 5 Traffic light phase information in case of dynamic time control

Because of the almost unlimited range of transmission over Internet and DAB Broadcasting, a TPEG-TSI message about an intersection may be received by vehicles far from this intersection.

TPEG-TSI provides more comprehensive information of traffic light compared to SPAT as illustrated in Figs. 4 and 5.

4 Presentation of Traffic Light Information on eHorizon via ADASIS v2

4.1 The Basic Idea

The solution presented in this section is based on a patent application submitted by Continental Automotive GmbH to the Deutsches Patent- und Markenamt on October 15, 2014 [7]. In our concept, we suppose that traffic light information of an intersection ahead of the ego-vehicle is available in 2 different ways,

(a) Only the actual state, e.g. red, green and amber as well as the rest duration of the current state are known. This is the case when the traffic light information is transmitted directly over short range communication, e.g. as C2X SPAT. Due to the short distance between the vehicle and the traffic light, the actual signal state and its rest duration are mostly interested.
(b) The complete signal phases, i.e. the duration of the green, amber and red phases are known. This is the case when the traffic light information is transmitted via TPEG-TSI. In this case, the vehicle can receive traffic light information of an intersection which is far from the ego position.

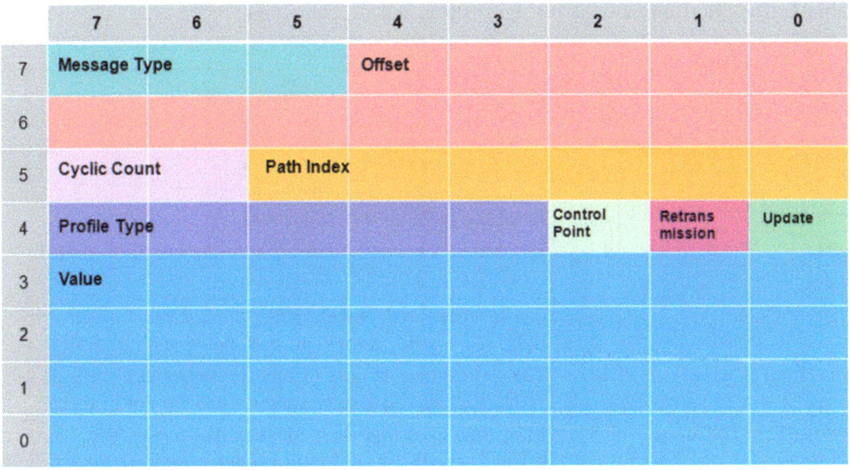

Fig. 6 Layout of the eHorizon message "PROFILE-LONG" [2]

The basic idea of the dynamic extension of eHorizon is to present the traffic light information on the eHorizon paths using ADAIS v2 interfaces.

We choose the PROFILE-LONG message [2] as container for the traffic light information, since it has the greatest payload (32 bit) among all eHorizon messages. The PROFILE-LONG message has the CAN-layout depicted in Fig. 6.

4.2 Construction of New eHorizon Message for Traffic Light Information

The eHorizon provides the road geometry as paths (sub-paths) and stubs. Using the road geometry data, the transmitted intersection and, particularly, stop lines can be resolved along with the eHorizon as following.

(a) Each stop line, for which traffic light information is available, will be matched to the corresponding stub on the eHorizon
(b) The stop lines on the MPP are referred as relevant stop lines
(c) For each relevant stop line, eHorizon-messages will be created for the traffic light timing information.

4.2.1 The eHorizon-Message for the Case that Only the Actual Signal Status and the Switching Time Are Known

In this case, e.g. C2X.message "Signal Phase and Timing" (SPAT), the signal status, the most likely as well as the earliest and the latest time for the next switch are known. The eHorizon-Message can then be defined as following Table 1.

Table 1 The eHorizon message for signal status and switching time

Tag	Bit	Content and format
Head		
Message Type	3	
Profile Type	5	=16
Cyclic Count	2	
Path Index	6	ID of the path, on which the stop line is
Offset	13	Position of the stop line as offset to the origin of the path
Value		
Current Color	3	0–7 for green, red, amber, amber-blinker, red + green arrow right, red + green arrow left, dark green and not available
MinTimeToChange	10	The earliest time of next switch; Resolution 0.1 s; Range 0–1021; 1022 for all greater values; 1023 for not available
IntervalTimeToChange	5	Max. time to change after MinTimeToChange; Resolution 0.1 s; Range 0–29; 30 for all greater values; 31 for not available
LikelyTimeToChange	5	Likely time to change after MinTimeToChange; Resolution 0.1 s; Range 0–29; 30 for all greater values; 31 for not available
Confidence	4	Probability of LikelyTimeToChange; Range 0–15; The value 15 means 100 %
GreenWaveSpeed	5	Reference speed for remaining in green wave; Optional; Range 0–30 m/s (0–108 km/h); 31 for not available

4.2.2 The eHorizon-Messages for the Case that the Signal Phase Is Known

In case of a fixed time control, the traffic light phase data like start of the phase, duration of the phase and the duration of a cycle, etc. are available.

In case of a dynamic time control, the traffic light phase data are available, but subjected to sudden changes. They can only be estimated within a band. According to the estimation confidences, additional values are provided as earliest start, most likely start, most likely end and latest end.

The key issue of the presentation of traffic light phase with eHorizon messages is the time factor which was so far not treated by the ADASIS v2 protocol. According to Fig. 6, the maximal effective payload of a PROFILE-LONG message is 32 bit and this is not sufficient for a standard time object. On the other hand, the traffic light phase must be attached with an unambiguous time indicator because of the non-synchronized message transmission in CAN.

To solve this conflict, we propose to attach one absolute time within a day (24 h) to the eHorizon-message and all other time variables are presented relatively to the absolute time.

Using the above traffic light data, 2 eHorizon-Messages are defined. The first message is for the traffic light phase without dynamic adaptation, i.e. in fact fixed time control.

The duration of the amber phase depends on the speed-limit and lasts usually a few seconds. In this paper, the amber phase is not explicitly treated and we define the non-green phase covering the red and amber phase, as shown in Table 2.

The second message will be created only, when the signal phase is not fixed time controlled (Table 3).

Table 2 The 1st eHorizon message for traffic light without dynamic adaption

Tag	Bit	Content and format
Head		
Message Type	3	
Profile Type	5	=17
Cyclic Count	2	
Path Index	6	ID of the path, on which the stop line is
Offset	13	Position of the stop line as offset to the origin of the path
Value		
ControlStatus	1	0 = fixed time; 1 = Dynamic time
NextStartGreen	17	Start of the next green phase within a day (24 h); Resolution 1 s
GreenPhase	7	Range 0–125 s; Duration of the green phase; 126 for values ≥126; 127 for not available
NoGreenPhase	7	0–125 s; Duration of the non-green phase; 126 for values ≥126; 127 for not available

Table 3 The 2nd eHorizon message for the case of dynamic time control

Tag	Bit	Content and format
Head		
Message Type	3	
Profile Type	5	=18
Cyclic Count	2	
Path Index	6	ID of the path, on which the stop line is
Offset	13	Position of the stop line as offset to the origin of the path
Value		
SignalDirection	4	Bit string
MostLikelyStart	7	Most likely start as delta before NextStartGreen; Resolution 0.1 s
MostLikelyEnd	7	Most likely end as delta after NextStartGreen + GreenPhase; Resolution 0.1 s
EarliestStart	7	Earliest start as delta before NextStartGreen; Resolution 0.1 s
LatestEnd	7	Latest end as delta after NextStartGreen + GreenPhase; Resolution 0.1 s

4.2.3 Multiple Messages in Sequence

According to the specifications, it is possible to present successive changes of traffic information for each stop line in just one SPAT or TPEG-TSI message. In this case, multiple eHorizon messages as described in Sects. 4.2.1 and 4.2.2 will be consequently created.

Acknowledgment The work presented in this paper was conducted within the research project URBAN of German automobile industry together with ICT companies, research institutes, universities and German cities. URBAN is partially funded by the Federal Ministry of Economics and Technologies of Germany. The author thanks for the financial support and, especially, for the constructive cooperation with the partners of the URBAN sub-projects "Smart Intersections" and "Urban Roads".

References

1. Scania Group, Scania Active Prediction (2011) Document downloadable from http://www.scania.com/products-services/trucks/safety-driver-support/driver-support-systems/active-prediction/. Dec 2011
2. ADASIS Forum, ADASIS v2 Protocol (2013) Document can be requested under www.adasis.ertico.com. Dec 2013
3. Ress C, Balzer D, Bracht A, Durekovic S, Loewenau J (2008) ADASIS protocol for advanced In-vehicle applications. In: 15th ITS World Congress. New York, 16–20 Nov 2008
4. Schweiger B, Bechler M (2011) Data processing in a vehicle, publication of world intelligent property organization WO 2011/117141 A1, 29 Sept 2011
5. Mortara P, De Gennaro M (2013) Method for planning the route of a vehicle, publication of european patent office. EP 2 610 782 A1, 3 July 2013
6. Kuefer J, Hudecek J, Kotte J, Zlocki A, Eckstein L (2014) A extended concept for a central information platform for perceptive, intelligent vehicle and its possibilities; automotive meets electronics. Dortmund, Germany, 18–19 Feb 2014
7. Pu H, Verfahren zur Fahrerassistenz unter Berücksichtigung einer Signalanlage, Patent application submitted to Deutsches Patent- und Markenamt DE 10 2014 220 935.8, 15 Oct 2014, Expected publication in April 2016

Virtual Stochastic Testing of Advanced Driver Assistance Systems

Stephanie Prialé Olivares, Nikolaus Rebernik, Arno Eichberger and Ernst Stadlober

Abstract With Advanced Driver Assistance Systems becoming increasingly complex, testing methods must keep up to efficiently test and validate these systems. This paper focuses on a method of testing vision-based Advanced Driver Assistance Systems on a state-of-the-art hardware-in-the-loop test bench. Virtual driving scenarios are being used for functional testing. This paper suggests a framework where the driving scenarios are constructed using a stochastical approach. This allows the testing of the parameter combinations that might otherwise be forgotten or disregarded by a human creating the scenarios. The first step of this framework, a road generator, is introduced. Generic courses of roads are created using the Markov Chain and Markov Chain Monte Carlo methods reconstructing real-life scenarios by analyzing map data.

Keywords Markov Chain · Markov Chain Monte Carlo · Stochastic road generator · Virtual driving scenario generator

S. Prialé Olivares (✉)
BMW Group and Institute for Real-Time Computer Systems, Technical University of Munich, Grillparzerstraße 25, 81675 Munich, Germany
e-mail: stephanie.priale@gmail.com

N. Rebernik
BMW Group and Institute of Automotive Engineering, Graz University of Technology, Luisenstraße 45, 80333 Munich, Germany
e-mail: nikolaus.rebernik@gmail.com

A. Eichberger
Institute of Automotive Engineering, Graz University of Technology, Inffeldgasse 11/II, 8010 Graz, Austria
e-mail: arno.eichberger@tugraz.at

E. Stadlober
Institute of Statistics, Graz University of Technology, Kopernikusgasse 24/III, 8010 Graz, Austria
e-mail: e.stadlober@tugraz.at

T. Schulze et al. (eds.), *Advanced Microsystems for Automotive Applications 2015*, Lecture Notes in Mobility, DOI 10.1007/978-3-319-20855-8_3

1 Introduction

As Advanced Driver Assistance Systems (ADAS) are becoming more and more complex, testing and validation processes and methods must keep up. The automation of driving functions is key for future mobility, facilitating greener cars with reduced fuel consumption, as well as improved road safety, by the minimization of hazardous situations in assisting the driver to react more quickly. However, automation of safety critical driving functions requires a minimum risk of failures, as described by norm ISO 26262. According to literature, a valid validation of automated driving functions in compliance with ISO 26262 demands a very large amount of testing kilometers [1], which will make the system expensive and hinder market penetration.

As a step towards a more efficient validation process, this paper looks at a new approach for subsystem integration using a state-of-the-art Hardware-In-the-Loop (HIL) test bench with a focus on vision-based ADAS demonstrated by the example of a lane detection system. HIL testing is a vital step in subsystem integration. It offers a test platform in early development stages and is cost-efficient, since the development can be started earlier. It also helps reduce the number of prototype vehicles needed.

In HIL environments, a variety of approaches are being employed to test and validate ADAS today: A fixed set of scenarios is being used and recorded scenes are being reprocessed or transferred to virtual environments: for example, using recorded sensor data to automatically transfer and create a scenario [2]. Augmented reality, as proposed in [3], is yet another approach that is being used as a Vehicle-In-The-Loop (VEHIL) setup. These approaches are either labor intensive or cannot adequately cope with the high complexity of future systems. Until now, the designing principle of intervening systems is not to try to cover all possible scenarios where a system could help but, rather, to focus on situations where a system will prove beneficial [4]. This, however, will change with increasing automation in driving, where vehicles will have to cope with many situations without the help of their driver. When looking into the future towards autonomously driving vehicles, a fixed catalogue will have to include thousands of tests to cover a sheer infinite number of scenarios. Creating them by hand will take much time and labor. It is said to take 50×10^8 km [1] of road testing to reliably validate autonomous driving functions. With this high amount of required kilometers, more efficient methods of validation must be implemented along the entire validation process throughout the development stages.

The goal is to automatically create virtual driving scenarios for testing image processing functions of vision based ADAS with an optimized parameter distribution to focus on more critical parameter ranges. One approach is to use combinatorics for the creation of test scenarios, as discussed in [5]. Scenarios are created by calculating all possible combinations of parameters and using equivalency class formation to reduce the number of scenarios to be tested.

This paper, however, proposes a stochastical approach. Since HIL testing is usually done in real time, it is not economically reasonable to test millions of possible outcomes and scenarios. Thus, a selection process has to be implemented. The parameters' influence on ADAS is analyzed using a design of experiment (DOE) approach. By doing this, the influence of parameters and their ranges on the correct functionality of the image processing algorithms of ADAS are evaluated. The likelihood of occurrence of parameter values for generating virtual driving scenarios will be higher in ranges where faults are more likely to occur, thereby focusing on critical areas while still testing the entire parameter ranges. As will the described in Chap. 2, real world statistical data to form distribution functions of road characteristics is obtained by analyzing map data as a basis for road generation. These functions describing road characteristics are then influenced towards the obtained critical ranges. This will be done in a future step, as described in the conclusion. Virtual environments have the advantage of knowing their properties. In that sense, fundamental information is known to which measurements of the ADAS can be compared. To evaluate the large number of generated scenarios, an automatic evaluation process must also be implemented. By testing a significant amount, depending on the ADAS function to be tested, of scenarios a qualitative statement can be made about the correct functionality of an ADAS function. Thereby a fast overview about an ADAS system can be obtained during the development stages. The method will be tested on vision-based ADAS; more specifically, on a lane detection system as a base for expansion towards more complex functions like lane keeping and congestion assistance systems, which, again, are a base for automated driving functions.

To reach this goal, the presented research will implement and test a number of steps. The first step is a road generator, which is the focus of this paper. It is the basis for creating driving scenarios and to establish and test this framework. This road generator will be explained in more detail in the next chapter.

2 System Model

In this section, the system model in charge of generating stochastic scenarios is described. Stochastic scenarios are the representation of the road's static elements, traffic, and surrounding elements, which are modeled according to probabilities computed by statistical analysis using different types of statistical methods. These scenarios seek a good approximation to scenarios that attempt to reflect reality as precisely as possible so that the ADAS can be properly tested.

Since close-to-reality scenarios are being pursued, the stochastic scenarios are based on statistical analysis on routes which actually exist. Essential characteristics such as geometry, type, and number of lanes, among others, are deduced from routes selected in OpenStreetMap. The system model takes this information, develops a database, and produces a probability density function (pdf) or conditional probabilities, depending on the attribute, for each aspect that needs to be evaluated to generate a scenario. Pdfs and conditional probabilities are a part of the statistical methods, thereby creating a statistical model in charge of generating roads with specific attributes.

A road is designed according to probabilities computed by the statistical analysis using different types of statistical methods. The following two methods are used for the evaluation and generation of scenarios: the Metropolis algorithm and the Markov Chain.[1]

These algorithms are directly influenced by the route chosen in OpenStreetMap. They are also indirectly influenced by constraints added to the system inside these algorithms to ensure its right functionality. Each time a new route is selected as base for the statistical analysis in OpenStreetMap, it results in a new pdf. This models the probability of attributes being sampled in the statistical methods, in turn influencing the resulting scenarios. Meanwhile, the constraints can also impact the road features depending on how they are set.

2.1 Building Statistical Analysis

The building of a statistical analysis, or, more precisely, the creation of a Bayesian statistical analysis, usually requires some fundamental analysis steps [6]. However, some posterior distributions are too complex to be calculated analytically and they are too time-consuming to sample. Using Markov Chain Monte Carlo (MCMC) algorithms, such as the Metropolis Hastings algorithm, can be very helpful for tackling both of these problems.

The algorithms used for the statistical analysis of the scenario's attributes are chosen according to the degree of complexity by the calculation of the posterior distribution and its high dimensional sampling. Two methods have been selected. One of them is the Metropolis algorithm (the special case of the Metropolis Hastings algorithm for symmetrical distributions) and the other is the Markov Chain. The Metropolis algorithm is in charge of creating the road geometry. This algorithm generates a random walk of points distributed according to a target distribution [7]. In this case, the points are being distributed according to the pdf that is a result of the analysis of the curvatures of the selected OpenStreetMap road. Since the resulting distribution for the geometry factor requires high dimensional sampling and can be too time-consuming with conventional Monte Carlo algorithms or Markov Chains, a MCMC algorithm is selected. Further points regarding this decision on the statistical methods are explained in 2.2.

On the other hand, a Markov Chain is required for those attributes where the Markov property also has to be fulfilled, but where sampling is not that time-consuming (e.g. setting the number of lanes on the road) and where algorithms for high dimensional distributions are not required. There are also attributes that possess constant values or are user-defined-data and, therefore, no algorithm is needed for their analysis.

---

[1]"Markov Chain" is used as reference to indicate a Markov process with a finite number of states.

2.1.1 Metropolis Algorithm Construction Steps

This algorithm works according to the structure presented in Fig. 1. The algorithm works by choosing an initial position x(0) [Step 1], then a proposed move x* is generated from the proposal distribution [Step 2]. This move will be either accepted or rejected according to an acceptance criterion $A(x^{(i)}, x^*)$, where x(i) represents the last accepted move and x* represents the proposed move. After performing a number of steps, the Metropolis Hastings algorithm generates a number of points, which construct the Markov Chain. These points will be distributed according to the desired distribution. The proposal distribution in the algorithm impacts the acceptance criterion. Therefore, it should be chosen in such a manner that it covers the target distribution completely. However, it is also important that it converges quickly and effectively. For continuous components, the Gaussian distribution or heavier-tailed distributions, e.g. Student's t distributions with low degrees of freedom, are commonly used [8]. Both of them are symmetric, which makes the acceptance criterion easy to overcome and produces a quick convergence. Another requirement of the Metropolis algorithm is to specify the target distribution.

The target distribution is fixed and calculated according to the data collection of the feature being analyzed. The calculation in this model is done in 3 steps:

1. Calculation of histogram based on data exported from OpenStreetMap.
2. Estimation of the pdf using kernel density estimation.
3. Specification of the (target) probability distribution according to the calculated pdf.

Once all requirements are provided, the algorithm is run and the samples related to the curvature of the road are generated. Unlike the features calculated with Markov Chains, these samples are not taken directly as attributes for the objects implemented in the system, but they have to be recalculated so they can be part of the objects belonging originally to the geometry of the road. This is required, since the road geometry is defined by spiral, arc, and line elements. These attributes are created according to the curvature sampled by the algorithm.

Figure 2 is an example of the curvature pdf for the highways around the city of Munich, calculated via kernel estimation.

The target distribution results from taking the resulting function of the kernel density estimation and setting the input values of the function as undefined. These values will be set once the algorithm is running. These undefined values take the

Fig. 1 Metropolis algorithm

1. Initialise $x^{(0)}$
2. For i = 0 to N-1
   - Sample $u \sim U_{[0,1]}$
   - Sample $x^* \sim q(x^*|x^{(i)})$
   - If $u < A(x^{(i)}|x^*) = \min\{1, \frac{p(x^*)}{p(x^{(i)})}\}$

     $x^{(i+1)} = x^*$

   else

     $x^{(i+1)} = x^{(i)}$

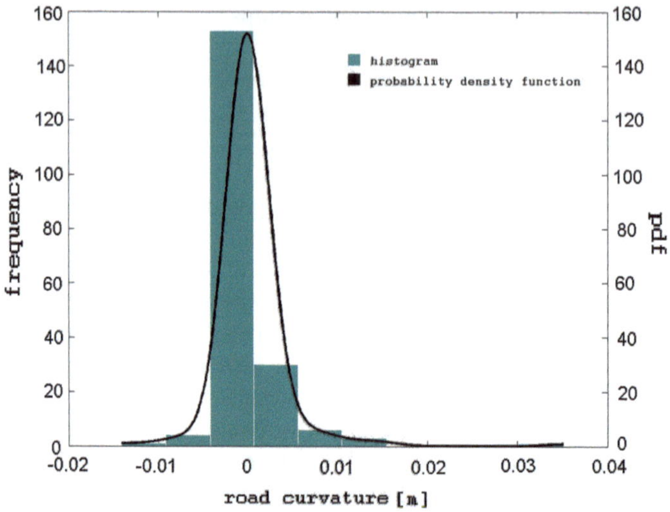

Fig. 2 Pdf estimation based on the highways around the city of Munich

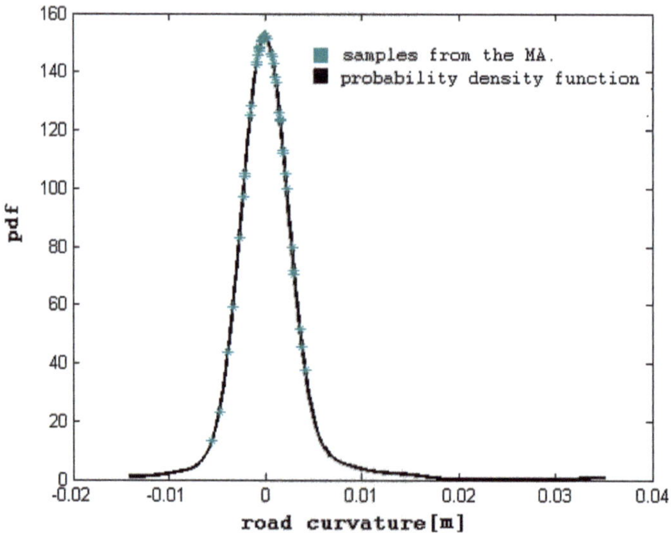

Fig. 3 Metropolis algorithm example for the highways around the city of Munich

sample values of the Markov Chain, which are built by the Metropolis algorithm (MA). Figure 3 shows an MA example for the highway around the city of Munich. This figure represents "x" values being sampled from the resulting pdf of the kernel estimation according to the Metropolis algorithm being associated with its corresponding pdf(x) value.

2.1.2 Markov Chain Construction Steps

Each Markov Chain is based on states that correspond to the possible values that the attribute being analyzed can take. Every time the Markov Chain is run, transitions between states occur according to probabilities given by the transition matrix P. These transition matrices are created according to the conditional probabilities of the attribute being analyzed, where each ijth entry represents a particular conditional probability: p (*moving to state* $j|$ *in state* i).

These conditional probabilities are calculated from the database of road features specified in OpenStreetMap. After setting the transition matrices, the stochastic row vector can be calculated through definition (1), where P represents the transition matrix and $x^{(n)}$ the stochastic row vector at step time n [6], and samples can be drawn. These resulting samples represent the attributes of the stochastic scenario, where the Markov property is fulfilled but where the sampling does not consume a great amount of time.

$$x^{(n+1)} = x^{(n)}\mathrm{P} \tag{1}$$

2.2 *Statistical Methods*

The objective of statistical methods is to make the process of scenario generation as efficient and productive as possible; hence, a proper selection of statistical methods is necessary. The most important part of choosing the correct test is to ensure that the test is appropriate for the type of data that has been collected [9], to observe adequate distributions of the variables, and to reflect what kind of relationship exists between them. The data collected for the system model is made up of the features taken from the route(s) chosen in OpenStreetMap, which are, in turn, the attributes required for the scenario generation. Some of these feature variables are correlated, meaning that the probability of a next event can increase or decrease based on the current event. These variables can be seen in the form of states and can be described with the help of Markov Chains. This is because a Markov Chain is generated based only on the previous state making new states likely to be correlated with the preceding state.

Depending on the variable being analyzed, a Markov Chain can be easily constructed and sampled. This happens by calculating the conditional probabilities from the database of road features specified in OpenStreetMap and setting the transition matrices. For the construction of a suitable Markov Chain, every possible state should be specified. Therefore, a Markov Chain is useful if the number of states is countable.

However, features for the analysis of the geometry construction of the road exist where the probability distribution is difficult to calculate and to simulate due to the high number of parameters. This can be solved by using Markov Chain Monte Carlo (MCMC) algorithms.

There are different kinds of MCMC algorithms. Among them, the most commonly used are the Metropolis Hastings algorithm and the Gibbs sampler. In the system model, the Metropolis Hastings algorithm is used instead of the Gibbs sampler, since, with the Metropolis Hastings algorithm, it is required to know the full conditional distribution up to a constant [10], while, with the Gibbs sampler, it is necessary to provide the full conditional distributions of the model in closed form [11]. This has to take place; otherwise, it is not possible to use the Gibbs sampler. Hence, the Metropolis Hastings algorithm becomes a more suitable method for the model. In the system model, the proposal distributions are symmetric; therefore, the simpler Metropolis algorithm is used instead.

2.3 Constraints

For both statistical methods (the Markov Chain and the Metropolis algorithm), constraints have to be set in order to ensure the right function of the system model. Conditional requirements vary depending on the attribute being analyzed.

These requirements are included in the algorithm in charge of creating the attribute during the generation process of stochastic scenarios such as, for example, to avoid a transition from a two lane-road to a four-lane road to be part of the stochastic scenarios. This constraint is implemented as part of the Markov Chain of the driving lanes' attributes in order to disable this possibility.

3 Performance Study

Different types of scenarios are constructed by varying the chosen route in OpenStreetMap. To inspect the performance of the system model even closer, the number of samples in the algorithms and repercussions of constraints in the generation of stochastic scenarios are evaluated.

3.1 OpenStreetMap Route Selection

When a new route in OpenStreetMap is selected for evaluation in the system model, the inputs for the algorithms change; in other words, the *transition matrix* required for the construction of the Markov Chain and the *pdf* taken as target distribution for the Metropolis algorithm change. For example, if a route is selected, where rather few curves but more straight stretches of road are presented, a *pdf* similar to Fig. 2 is expected to occur. On such a route, the highest probability resides in the curvature with value 0, and the algorithm tends to sample towards this value. In the case of routes with a higher probability of curvature, resulting roads with strong curvatures have a higher probability to appear.

The *transition matrix* necessary for the calculation of the Markov Chain changes when new routes are selected. However, since some of the non- road's geometry features, such as road markings, are based on fixed road specifications, their probabilities do not vary among different selected routes.

3.2 Constraints

Constraints used in the algorithms for the correct functionality can affect the road's features in the stochastic scenario depending on how well they are tuned. In the case of the geometry, a maximum or minimum duration of a curvature can be established in order to generate stochastic scenarios that fulfill specific characteristics expected by the user.

For features such as speed limit, constraints can be set so that jumping between values is not possible but a linear increment is given. This is also a feature set according to user desire.

Figure 4 shows how varying the constraint set for the geometry attribute can influence the smoothness of the curve despite maintaining the same probability density. The three resulting roads represent different curvature constraints where the spiral in charge of creating the curvature uses less iteration steps on each figure from left to right. By using a lesser number of iterations to diminish the curvature, the curvature ends up having a higher tangent and a stronger curve.

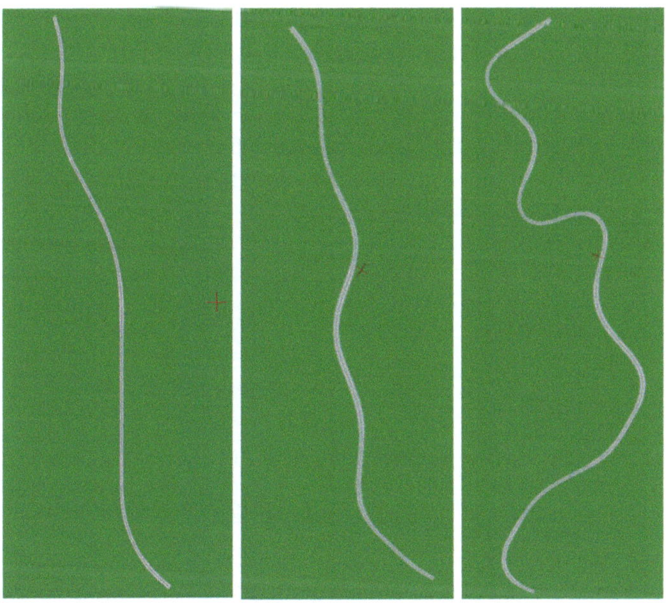

Fig. 4 Output roads with different curvature constraints

3.3 Number of Samples in Statistical Methods

The number of samples or steps the algorithms need to create each road attribute has an influence on the resulting roads. If the number of samples for the algorithms increases, the probability of sampling a particular value, which is normally low, increases as well. Furthermore, in case a constraint determines that an attribute needs to remain constant for a defined amount of iterations, these already low-probability values can reach an even lower probability if the number of iterations decreases, and vice versa.

4 Conclusion and Outlook

Using a stochastic approach by employing Markov Chain and Markov Chain Monte Carlo methods, and using real world data to develop the corresponding probability density functions for road-generating algorithms, generic but still realistic roads can be created. The real world data is being obtained by analyzing roads exported from OpenStreetMap. The discussed road generator is the basis for generating virtual driving scenarios and for the suggested framework.

The next steps will include a design of experiment approach to optimize the probability density functions towards critical parameter ranges. With this approach the main influencing parameters on the image processing functions of the ADAS, concerning the road creation, will be evaluated. The critical parameters as well as their critical value ranges, which cause the most failures of the vision based ADAS, are of interest. The pdf's of the road generator will be influenced towards these more critical ranges. Another step is the expansion of the road generator to include different road markings, roadside structures as well as construction zones. Construction zones are challenging for lane detection functions as there are typically many different markings and structures on the road that make detecting the correct driving lane more difficult.

This framework contributes a further approach to tackle the challenges of validating complex ADAS with an outlook on automated driving functions.

References

1. Winner H (2013) Absicherung automatischen Fahrens, 6. FAS-Tagung München, Munich
2. Lages U, Spencer M, Katz R (2013) Automatic scenario generation based on laserscanner reference data and advanced offline processing. In: Intelligent vehicles symposium workshops (IV workshops), pp 146, 148
3. Zofka MR, Kohlhaas R, Schamm T, Zöllner JM (2014) Semivirtual simulations for the evaluation of vision-based ADAS. In: Intelligent vehicles symposium proceedings, IEEE, pp 121, 126

4. Schwarz J (n.d.) Response 3—code of practice for development, validation and market introduction of ADAS—A PReVENT project. DaimlerChrysler AG, Stuttgart
5. Schuldt F, Sausts F, Lichte B, Maurer M (2013) Effiziente systematische Testgenerierung für Fahrerassistenzsysteme in virtuellen Umgebungen. In: AAET2013—Automatisierungssysteme, Assistenzsysteme und eingebettete Systeme für Transportmittel, Braunschweig
6. Gamerman D, Lopes H (2006) Markov Chain Monte Carlo: stochastic simulation for bayesian inference. Taylor & Francis Group, Boca Raton USA
7. Müller P (2009) Monte Carlo methods and bayesian computation: MCMC, vol 10
8. Neal R (1993) Probabilistic inference using Markov Chain Monte Carlo methods, U. Toronto
9. Vowler S (2007) Analysing data—choosing appropriate statistical methods. Hosp Pharmacist 44:12
10. Mengersen KL, Tweedie RL (1996) Rates of convergence of the hastings and metropolis algorithms. Ann Stat 24:101–121
11. Gilks W, Richardson S, Spiegelhalter D (1996) Markov Chain Monte Carlo in practice. Chapman & Hall/CRC, London

Shockwave Analysis on Motorways and Possibility of Damping by Autonomous Vehicles

Nassim Motamedidehkordi, Thomas Benz and Martin Margreiter

Abstract Shockwaves are a boundary that shows discontinuity in a flow-density domain. The physical realization of a shockwave is a point in time and space at which vehicles change their speed abruptly. The formation and dissolving of congestion are phenomena that are important for the traveler information and congestion management perspectives. Shockwave analysis is the method to identify congested areas and estimate the rate of formation and dissipation of the congestion. The microscopic traffic simulation tool *Vissim* was used to address the main objective of this study, namely to determine if and to what extent the driving behavior parameters of the model used influence the shockwaves on motorways. After precise calibration of the car following behavior based on the detected shockwaves from data of the German research project sim^{TD}, the possible influences on driver behavior through highly automated vehicles was sketched in order to figure out whether these applications can change the shockwave propagation speed on motorways, lead to suppression of shockwaves and improve the network performance as well as increase the traffic safety.

Keywords Shockwave · Motorways · Autonomous driving

N. Motamedidehkordi (✉) · M. Margreiter
Traffic Engineering and Control, Technical University of Munich,
Arcisstraße 21, 80333 Munich, Germany
e-mail: nassim.motamedidehkordi@tum.de

M. Margreiter
e-mail: martin.margreiter@tum.de

T. Benz
PTV Group, Haid-und-Neu-Straße 15, 76131 Karlsruhe, Germany
e-mail: thomas.benz@ptvgroup.com

T. Schulze et al. (eds.), *Advanced Microsystems for Automotive Applications 2015*, Lecture Notes in Mobility, DOI 10.1007/978-3-319-20855-8_4

1 Introduction

On motorways basically two types of traffic jams can occur: jams with a fixed bottleneck location and jams with upstream moving head and tail. In this study we focus on the second type [1]. Shockwaves can be defined as a by-product of the congestion which is originated from a sudden, substantial change in the state of traffic flow. They are the transition zones between two traffic states that move through the traffic like a propagation wave. A shockwave is also created by vehicles that accelerate after being released from a queue. The propagation can be backward or forward [2]. As travel time, air pollution and noise pollution for the vehicles entering the traffic jam increase and on the other hand the safety is reduced, studying the possibility of the damping of the shockwave is of high importance. Due to the fact that shockwaves usually have lower outflow than the capacity of the motorway, there might be a potential improvement of the traffic flow by means of suppression of the shockwave [3]. Shockwave theory describes how the boundary between two traffic states propagates through time and space [4].

The shockwave theory used in this paper is the one proposed by Lighthill and Whitham [5]; shockwave theory is described briefly here. The propagation speed of the boundary between traffic states can be visualized using the fundamental diagram of traffic flow, which explains the interdependencies between traffic volume, traffic density and mean speed. The slope of the line which connects the traffic states A and B in the fundamental diagram shown in Fig. 1) equals the propagation speed of the boundary between both traffic states. However, Lighthill and Whitham believed that the front between the two states in the left figure has the same slope of the line in right figure.

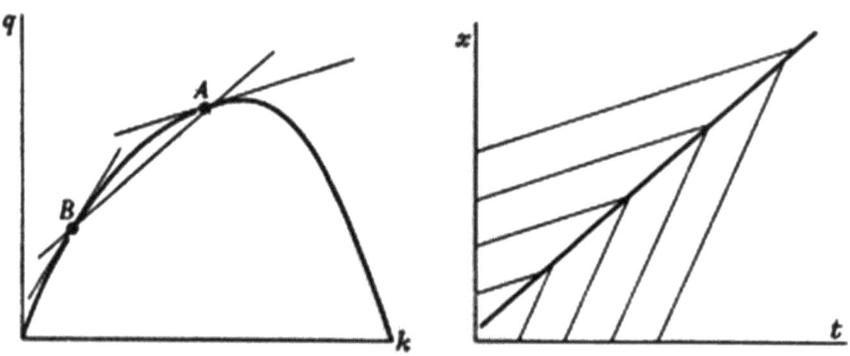

Fig. 1 The shockwave is shown as a *bold line* in space-time diagram on the *right*. Behind it the density is less and the waves travel faster. Ahead of it the flow is denser (*A*, *B* two different traffic states, *q* traffic volume, *k* traffic density, *x* distance, *t* time) [5]

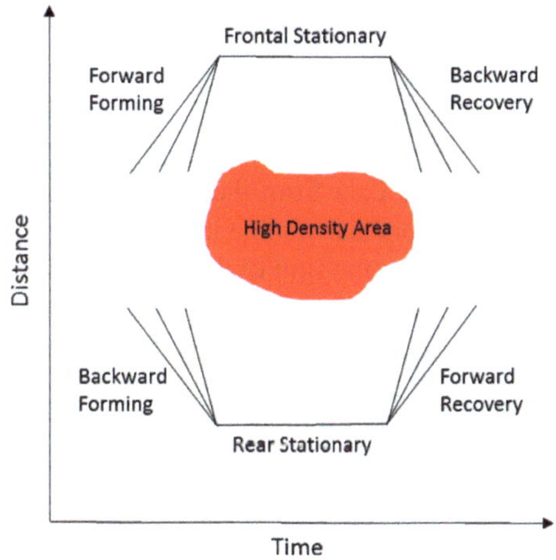

Fig. 2 Classification of shockwaves according to Hoogendoorn [14]

According to May [6] shockwaves can be classified into different categories (see Fig. 2):

1. Forward Forming (the congestion is gradually extending farther to the sections downstream backward forming).
2. Frontal Stationary (can be present at a bottleneck location and indicates the location where traffic demand exceeds capacity).
3. Rear Stationary (happens when the arriving traffic demand is equal to the flow in the congested area for some period of time).
4. Forward Recovery (occurs when there has been congestion, but demands are decreasing below the bottleneck capacity and the length of the congestion is being reduced).
5. Backward Recovery (occurs when there has been congestion, but demands are decreasing below the bottleneck capacity and the length of the congestion is being reduced).

Dynamic speed limits and communication by road side systems have proven to be successful instruments in stabilizing the traffic flow [7] and in dissolving the shockwave [8]. This study examines whether or not autonomous driving can lead to the suppression of shockwaves.

This article is organized as follows: in Sect. 2 we describe the empirical data as well as the methodology for data analysis. Simulation and calibration steps of the *Vissim* network based on the empirical data are performed in Sect. 3. In Sect. 4 we examine the effect of fully automated vehicles on transient behavior of traffic flow and shockwave propagation at the macroscopic level when they operate together with manually driven vehicles. Section 5 contains conclusions derived from our analysis and recommendations for further research.

2 Empirical Data and Methodology of Data Study

There are two sources of real empirical data which have been measured on the same days and road sections and were used for this study. We study the microscopic and macroscopic data from the German sim^{TD} project. Macroscopic data measured by loop detectors is available in aggregations of one minute. The variables recorded for each loop are: the sum of flows and speeds over lanes for Heavy Good Vehicles and Passenger cars as well as the concentration which is calculated based on flow and speed. Detailed sub-second vehicle probe data from the VIS, RIS and communication between the vehicles were used to reconstruct the vehicle trajectories and derive the speed profiles along these trajectories.

The motorway segment is an 8.5 km long section of motorway A5 in Hessia, Germany between Friedberg and Bad Homburger Kreuz in the North of

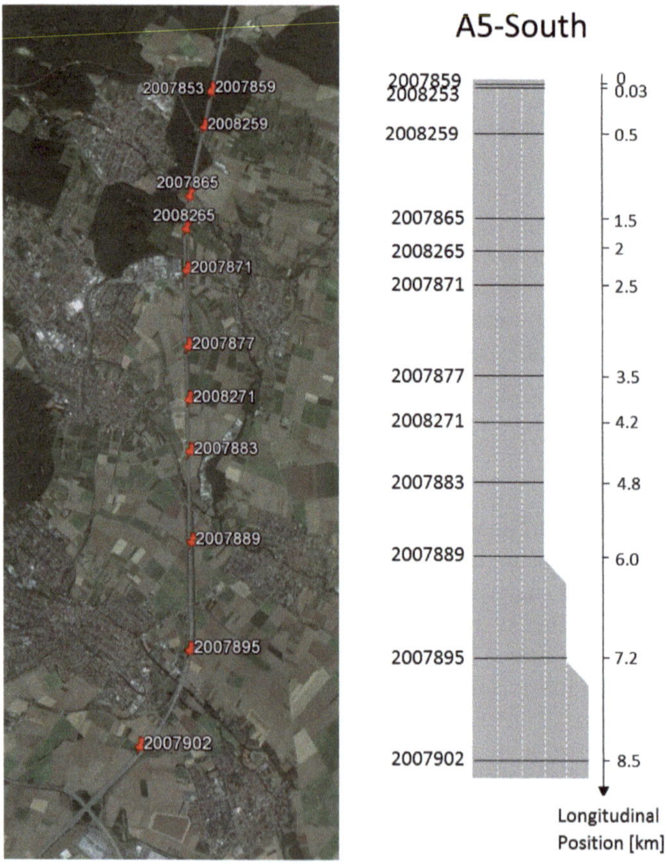

Fig. 3 The considered motorway stretch: a part of the German A5 motorway from Friedberg to Bad Homburger Kreuz

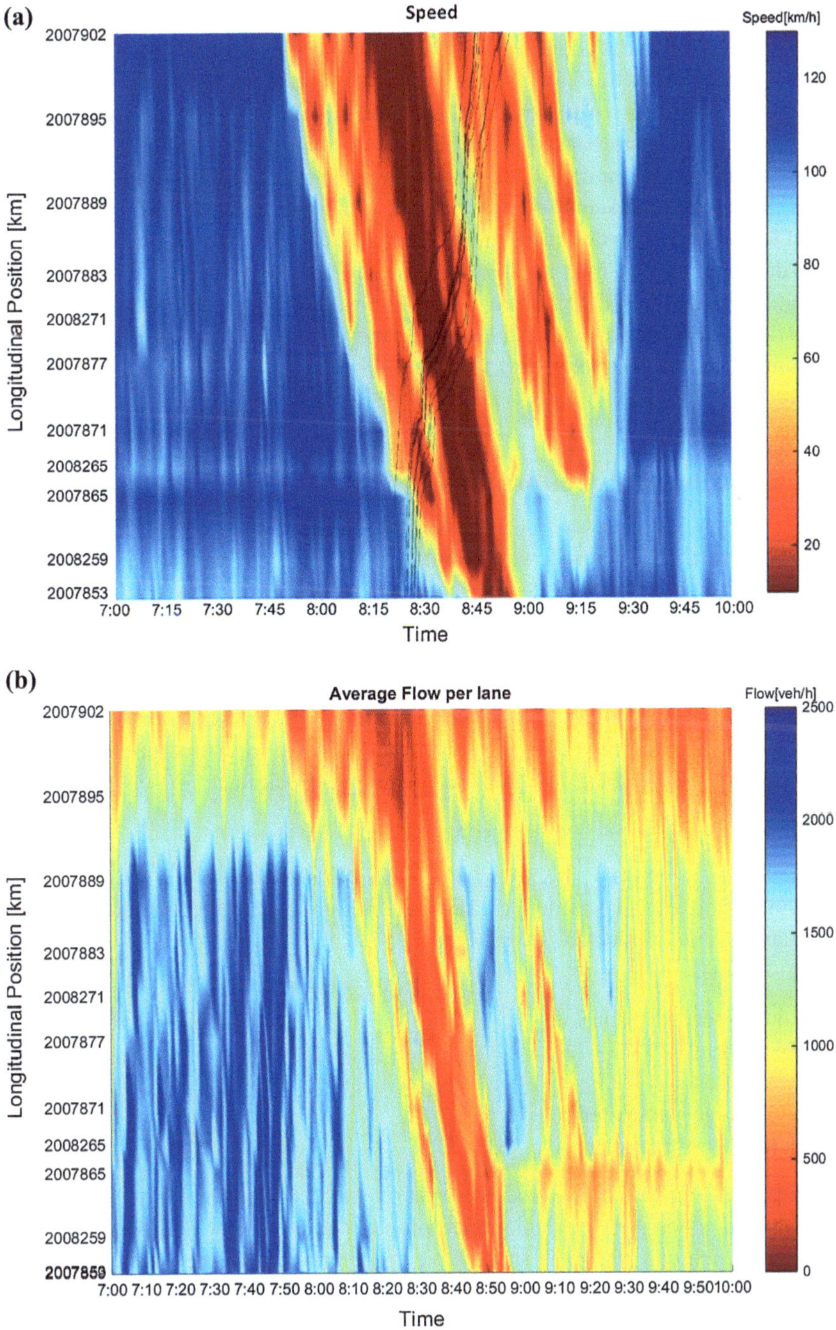

Fig. 4 Space-time diagrams of the **a** speeds and **b** flows derived from empirical data. The shockwave is observed and it is propagating backwards

Frankfurt/Main. We chose this section because congested traffic is observed there on 2012-10-11 in the data coming from 12 sets of loops detectors on this motorway segment and vehicle trajectories. This congestion is a major congestion which existed for a long time. Additionally the segment is limited between two ramps in order to avoid the disturbances of traffic flow due to merging traffic.

The result of the empirical data can be found in the graphs in Fig. 4. In these graphs the x axis represents the time and the y axis the longitudinal location with respect to the road. The detector numbers are also shown in the graph. The location of the detectors can be found in Fig. 3.

From comparison of detector data and vehicles trajectories, we can conclude that the spatiotemporal structures of the traffic pattern measured through two different data sources are qualitatively very similar.

3 Simulation and Calibration

For simulation the *Vissim* software is used. In *Vissim* there are various parameters that can be adjusted in order to replicate the reality. The tasks which should be completed in model setup are choosing the physical network, collection of field data and traffic compositions. As the physical network, a part of the network applied for the sim^{TD} project was used.

The first step to develop a model which represents the German motorway is to get the simulation in which the vehicle fleet matches the measured fleet. Therefore, based on the detector data, two vehicle classes were defined:

- Class 1: Heavy Good Vehicles (HGV).
- Class 2: Passenger Cars.

A simulation model should be calibrated before any further simulation-based traffic analysis is conducted. In this study, the calibration is focused on the following settings:

- Desired speed distribution
- Vehicle fleet
- Relative flows
- Lane change parameters
- Car following parameters

The parameters like the desired acceleration and deceleration distribution cannot be calibrated since the available microscopic data was not sufficient. Another parameter which can affect the shockwave speed and size of the shockwave is the driver's reaction time. This parameter is missing in most of the traffic simulation models. In *Vissim*, some implicit reaction time is modeled inherently by the action threshold of the psycho-physical model which is relevant only if the driver is changing the driving regime, which means changing the traffic state. But within one regime or traffic state the reaction time is not modeled and the simulation time step is interpreted as a reaction time which is a constant value over time for all the drivers and not realistic [9].

3.1 Model Setup

Desired Speed Distribution The distribution for desired speeds of passenger cars and HGVs were estimated based on the recorded data in low volume periods.

Vehicle Fleet Vehicle fleets and the compositions, for one minute interval, were derived directly from the detector data on the border of the network from 7:00 to 10:00 a.m.

Lane Change Parameters The calibration of lane change parameters for the drivers' behavior requires more detailed microscopic data like floating vehicle data or trajectories extracted from the video cameras. Hence, the lane change parameters in *Vissim* based on the model of Sparmann [10] were kept as default.

3.2 Calibration

In order to calibrate the car following parameters, it has been decided that no automatic method like genetic algorithm or heuristic method, is going to be used. In Wiedemann 74 [11], the existence of parameter z and its statistical spread results in variations between the driving behavior of the drivers and the distribution of the safety distance (Fig. 5).

In this model the minimum desired following distance between the vehicles is calculated as

$$ABX = AX + (bx_{add} + bx_{mult} * z) * \sqrt{v}$$

v Speed of the slower vehicle [m/s]

z Is a value of range [0,1] which is normally distributed around 0.5 with a standard deviation of 0.15

AX Average standstill distance which defines the average desired distance between two cars [m]

bx_{add} Additive part of safety distance which allows to adjust the time requirement values [−]

bx_{mult} Multiplicative part of safety distance which allows to adjust the time requirement values [−]

Fig. 5 Car following logic and driving states [15]

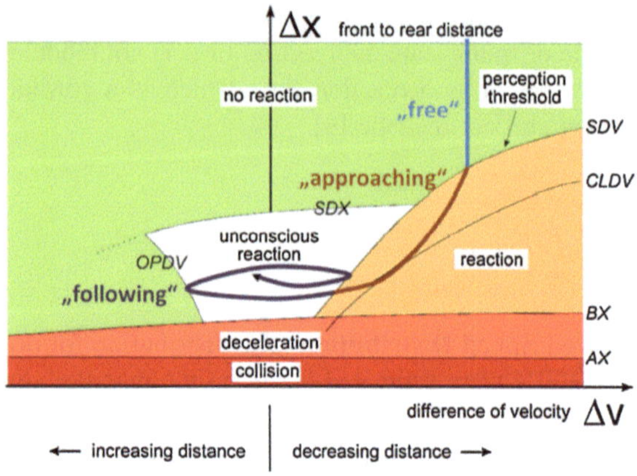

In order to keep the number of scenarios investigated for calibration limited and as the majority of vehicles are passenger cars, the calibration focuses on parameterization of passenger cars only. Parameter ranges were modified on the basis of field speed data and calculated minimum safety distance. The updated parameter set and its ranges are as follows:

- Average standstill distance = a value in range of 1–4
- Additive part of safety distance = a value in range of 0–8
- Multiple part of safety distance = a value in range of 1–8

Among the generated parameter set, 161 combinations of the three parameters above are chosen based on the engineering judgment and three different simulation runs were conducted for each set. After evaluation of the simulation the parameter which minimized the sum of the Root-Mean-Square Deviation (RMSD) of all measurement points over the three simulation runs was chosen as a best fitting parameter. RMSD for each detector calculated based on the formula below:

$$RMSD = \frac{\sqrt{\sum_{t=1}^{180}\left(\text{speed}_t^{Real\ Data} - \text{speed}_t^{Simulation}\right)}}{180}$$

The spatiotemporal graph of speed and flow of this parameter sets were used as a base scenario for analysis on this study. In Fig. 6 spatio-temporal diagrams of the speed and flow for the best combination set is found.

This driving behavior parameter set is used as a base driving behavior for all the manually driven vehicles in further simulations.

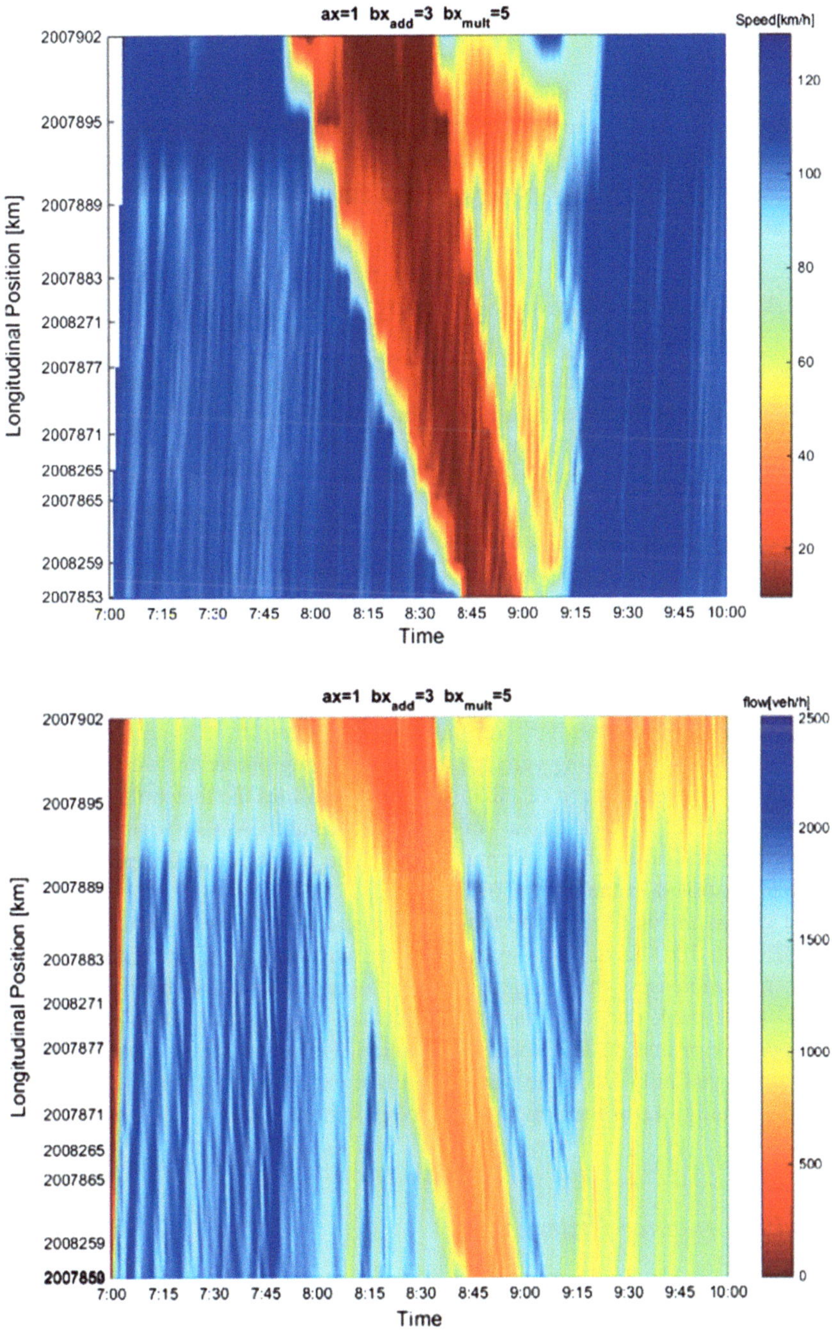

Fig. 6 Space-time diagram of the speed and flow resulted from calibration

4 Applications and Evaluation

In this section we present how autonomous driving can influence the traffic flow and the propagation of a shockwave and compare this to the base scenario. The base scenario was defined by using the parameters derived from calibration process. The simulation result for the base scenario can be found in the Appendix.

4.1 Autonomous Driving Scenario

Our scenario considered autonomous driving along with manually driven vehicles. These vehicles use the Wiedemann 99 following behavior model since they drive more homogeneously and there is not a significant variation in their driving behavior. To reduce the congestion and seemingly unexplained traffic jam it is important to maintain steady state flow between string of vehicles which can be achieved by uniform distribution and similar constant velocities [12]. The desired speed distributions of these vehicles were changed to more homogenous speeds while maintaining the same average speed as the manually driven vehicles. The relevance of achieving small headways becomes clear only when considered against the highway policy governing safe vehicle spacing. Current regulations allow the average driver to interpret and react to potential hazards quickly enough to avoid collisions. Sensitivity of the driver reaction to the acceleration and deceleration of the preceding vehicles during following procedure were set to −0.1 and +0.1 respectively instead of −0.35 and +0.35. Besides the threshold for entering following was set to −11 instead of −8. This scenario was also simulated with penetration rates of 5, 10, 20, 50 and 100 %. The results for different penetration rates are illustrated in the graphs of Fig. 7.

The graphs in Fig. 7 illustrate the resulting speed and flow diagrams over time (from 7:00 to 10:00 a.m.) and space for different penetration rates. As the penetration rate increases, the congestion area becomes less and free flow speeds become more stable due to the fact that automated vehicles have more homogenous driving behavior and desired speed distribution. However, the noticeable changes can be observed only when the penetration rate exceeds 20 %. On the other hand the flow coming out of the congestion becomes more.

4.2 Performance Criterion

The performance criterion for each scenario was the average network speed and the propagation speed and resolving speed of congestion along the motorway. The average network speed was calculated based on the formula:

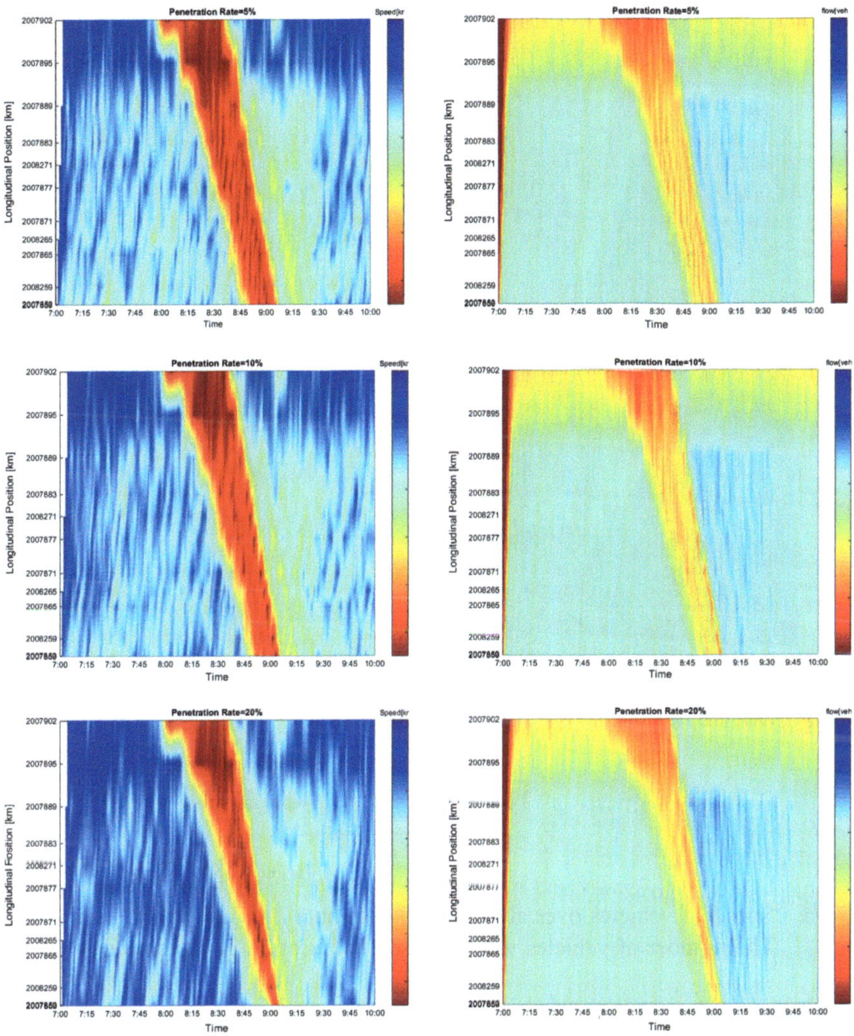

Fig. 7 Space-time diagrams of the speed (*left*) and flow (*right*) for autonomous driving with different penetration rates

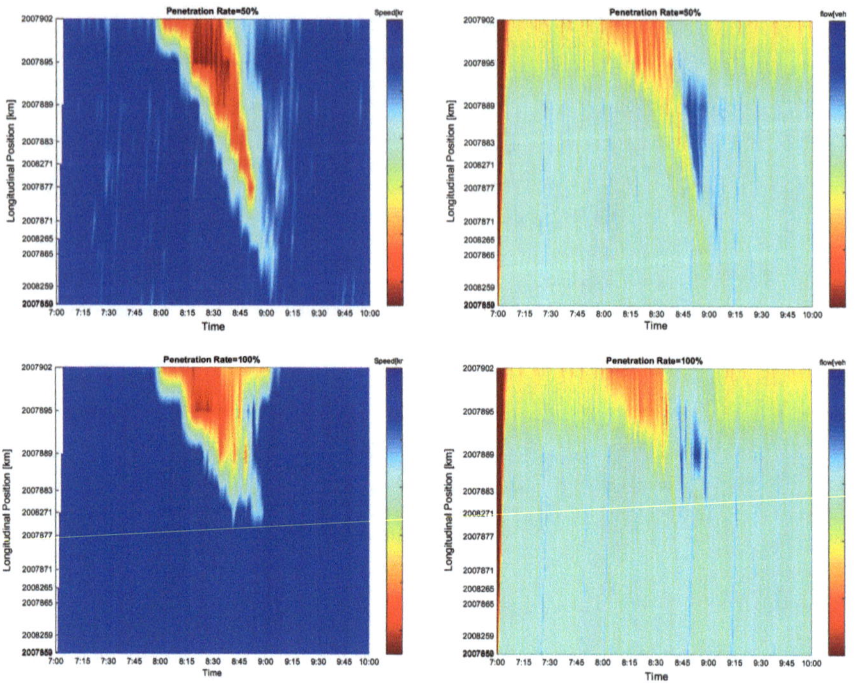

Fig. 7 (continued)

$$V = \frac{\sum_{i=1}^{180} Speed_i * flow_i}{\sum_{i=1}^{180} flow_i}$$

$speed_i$ Speed of vehicles over all lanes in ith minute
$flow_i$ The number of vehicles which passed that section in ith minute

In the scenario it seems that the 20 % penetration rate is enough to influence other vehicles noticeably. In case of fully automated vehicles the network performance improved significantly as we reduced the safety distance between the vehicles and thus increased the capacity of the motorway (Table 1).

According to the shockwave theory the propagation of the front between two traffic states has the same slope as the line corresponding the two states in the density-flow diagram [12]. Therefore instead of using q and k to calculate the propagation speed of the shockwave, the sudden change of speed in the space-time

Table 1 Average network performance for different penetration rates

Penetration rate (%)	Average speed (km/h)	Percentage change (%)
0	83.22	–
5	84.10	1.1
10	85.92	3.2
20	90.76	9.1
50	105.60	26.9
100	112.30	34.9

diagram was considered. The border of the jam at each detector was chosen as traffic state change from free flow to the speeds below 40 km/h [13]. Then simple linear regression was used, which would provide the best fit to the data points, in order to get the shockwave propagation speed of each scenario. The slope of the line is calculated based on following minimization problem using least squares.

$$\mathit{Find}\ \min Q(\alpha, \beta),$$
$$\mathit{for}\ Q(\alpha, \beta)\varepsilon_i = \sum_{i=1}^{n} \hat{\varepsilon}_i^2 = \sum_{i=1}^{n} (y_i - \alpha - \beta x_i)^2$$

The resulting shockwave speeds in each scenario are illustrated in the Table 2. The values are negative since the shockwave was propagating backwards:

In the scenario the shockwave propagates backwards slower as the penetration rate increases. For the shockwave of congestion dissolving the same methodology was used and the sudden increase in speeds recorded by the detectors to the values above 40 km/h was considered as a border. The Table 3 illustrates the speed of the dissolving shockwave.

In contrast to the propagation speed of the shockwave, the resolving does not change significantly in the scenario as the penetration rate increases and it stays rather constant. At a penetration rate of the 100 % a forward recovery shockwave was observed. Hence we did not calculate the percentage of the change for this penetration rate.

Table 2 Shockwave propagation speeds for different penetration rates

Penetration rate (%)	Shockwave propagation speed (km/h)	Percentage change (%)
0	−11.17	–
5	−10.49	−6.1
10	−10.17	−9.0
20	−8.78	−21.4
50	−6.26	−44.0
100	−4.81	−56.9

Table 3 Resolving speed for different penetration rates

Penetration rate (%)	Shockwave resolving speed (km/h)	Percentage change (%)
0	−18.08	–
5	−17.2	−4.9
10	−17.28	−4.4
20	−17.44	−3.5
50	−18.2	0.7

5 Conclusion and Recommendations

In this paper we discussed applications of the autonomous vehicles and the role that they play with regard to changing the propagation and dissolving of congestions on motorways. We have reported the result of applying autonomous driving along with manually driven vehicles in the microscopic traffic simulation *Vissim*. In the simulations a base scenario of three hours was used with shockwave propagation over a 8.5 km-long stretch on the German motorway A5.

The simulation results show that autonomous driving decreases the propagation speed of the shockwave and also increases the general network performance. It seems that the dissolving speed of the shockwave does not change when using automated vehicles.

Traffic jams are classified into two different kinds of jams [3]:

- Spontaneous jam or phantom jam which propagates backward as stop and go wave.
- Stationary jam which is induced by slowdown or blockage at a section of roadway.

In the flows near the capacity of the highway, if the sensitivity of the driver is lower than a critical value, the phantom jams occur. This study only focused on the stationary jam. Further investigation will address the suppression of phantom jams and the evaluation of the proposed measures for a wider range of scenarios. Finally, the security issues of the proposed research, especially autonomous driving, need to be studied carefully.

Appendix

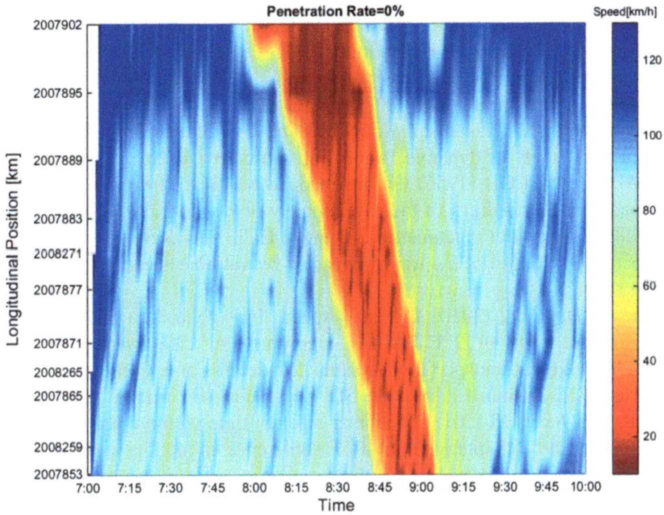

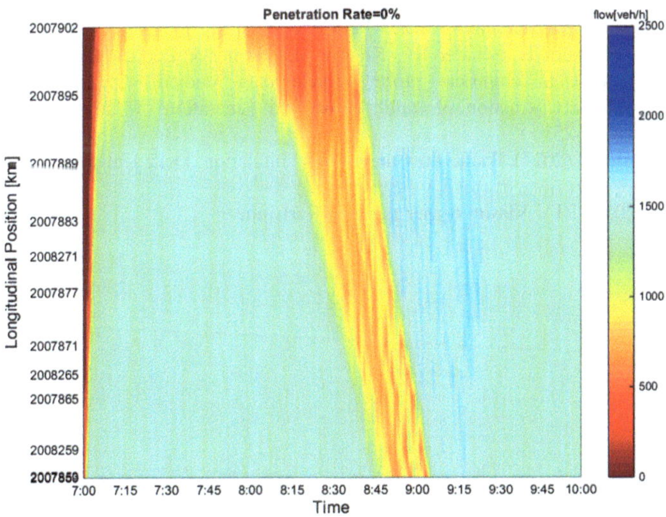

The space-time graph of speed and flow for the base scenario which includes 0 % automated vehicles.

References

1. Treiber M, Kesting A (2013) Traffic flow dynamics. Springer, Berlin
2. Tarko AP (2003) Highway traffic operations. Purdue University, CRC Press LLC
3. Hegyi A, Burger M, Schutter B, Hellendoorn J, van den Boom T (2007) Towards a practical application of model predecticve control to suppress shock waves on freeways. In: Proceeding of the European control conference 2007, Kos, Greece
4. Bando M, Hasebe K, Nakanishi K, Nakayama A, Shibata A, Sugiyama Y (1995) Phenomenological study of dynamical model of traffic flow. J Phys I 5(11):1389–1399
5. Lighthill MJ, Whitham GB (1956) On kinematic waves II: a theory of traffic flow on long crowded roads. In: Proceedings of the royal society of London, pp 317–345
6. May A (1990) Traffic flow fundamentals. Prentice-Hall, New Jersey
7. Smulders S (1990) Control of freeway traffic flow by variable speed signs. Trans Res Part B Methodol 111–132
8. Hoogendorn S, Hegyi A (2010) Dynamic speed limit control to resolve shockwaves on freeways-field test result of SPECIALIST algorithm. In: 13th international IEEE, annual conference of intelligent transportation systems, Madeira Island, Portugal
9. Basak K, Hetu SN, Zhemin L, Azevedo CL, Loganathan H, Toledo T, Runmin X, Yan X, Li-Shiuan P, Ben-Akiva M (2013) Modeling reaction time within a traffic simulation model. In: Proceedings of the 16th international IEEE conference on intelligent transportation systems (ITSC 2013)
10. Sparmann U (1978) Spurwechselvorgänge auf Zweispurigen BAB-Richtungsfahrbahnen. Forschung Straßenbau und Straßenverkehrstechnik, Heft 263, Bonn
11. Wiedemann R (1974) Simulation des Straßenverkehrsflusses, Schriftenreihe des Instituts für Verkehrswesen der Universität Karlsruhe, Heft 8
12. Hanaura H, Nagatani T, Tanaka K (2007) Jam formation in traffic flow on a highway with some slowdown sections. Phys A 374:419–430
13. Kerner BS, Rehborn H, Schäfer RP, Klenov SL, Palmer J, Lorkowski S, Witte N (2013) Traffic dynamics in empirical probe vehicle data studied with three-phase theory: spatiotemporal reconstruction of traffic phase and generation of Jam warning messages. Phys A 391:221–251
14. Hoogendoorn SP (2014) Traffic flow theory and simulation. Delft University of Technology, Delft
15. PTV AG (2013) PTV Vissim 6 user manual. Karlsruhe

Part II
Advanced Sensing Concepts

Driver Head Pose Estimation by Regression

Yodit Tessema, Matthias Höffken and Ulrich Kreßel

Abstract Advanced driver assistance systems provide a significant improvement for road safety and effective driving. Integrating driver observation in such systems is a crucial step in striving to reducing traffic accidents. For improved robustness in the designed system, we have chosen to approach the driver head pose estimation problem through regression methods based on images from a stereo camera. Therefore, the system operates solely on the 3D head information and is independent of facial feature detection, color and texture information. The proposed system contemplates real driving situations along with the design dictated position of the camera inside the car. The proposed regression algorithms for this work are support vector regression (SVR), random regression forest (RRF) and extremely randomized trees (ERT). Carried experimental studies show high accuracies for the proposed methods. Their algorithmic simplicity and measured time-costs further indicate their suitability for embedded real-time applications.

1 Introduction

The fatality rate and the economic cost resulting from inattentive driving prompted the design of robust Advanced Driver Assistance Systems (ADAS). Motivated by the desire to increase road safety, the systems are designed to assist the driver by monitoring the vehicle's surrounding environment and by providing essential

Y. Tessema (✉) · M. Höffken
Institute of Measurement, Control and Microtechnology, Ulm University,
Helmholtzstr. 18, 89081 Ulm, Germany
e-mail: yodit.tessema@uni-ulm.de; yodit.tessema@daimler.com

M. Höffken
e-mail: matthias.hoeffken@uni-ulm.de

Y. Tessema · U. Kreßel
Research and Development, Driver Analysis, Daimler AG, Wilhelm-Runge-Str. 11,
89081 Ulm, Germany
e-mail: ulrich.kressel@daimler.com

T. Schulze et al. (eds.), *Advanced Microsystems for Automotive Applications 2015*,
Lecture Notes in Mobility, DOI 10.1007/978-3-319-20855-8_5

information to the driver so that possible risks can be identified at an early stage. In dangerous situations the systems alarm the driver and, if required, take necessary measures automatically in an attempt to avoid accidents from occurring. To enhance the significance of such systems, their capability should be extended to monitor and analyze the activities of the driver.

In the endeavor to achieve autonomous driving, driver observation is even more inevitable. For the interaction of the system with the driver, it is necessary to ascertain whether the focus of the driver is on the driving process.

Observing head movement provides important information on the interest and intention of the driver. It is also an input for calculating the field of view which can in turn be used to approximate the eye gaze of the driver. Nonetheless, the extensive variation in individual appearance resulting from hairstyle, difference in facial features and head shape make accurate head pose estimation a very challenging task [16].

The framework of this paper is adopted from Höffken et al. [12], in which the preprocessing system chain is composed of a series of steps that the images pass through in order to extract only the relevant information needed for the task at hand. Although tracking improves the performance and robustness of the system, the final outcome is mainly dependent on the performance of the system on frame-wise estimation. Therefore, this work aims at improving the accuracy of the frame-wise head pose estimation. The integration of tracking will be considered as a future extension.

In order to increase the robustness of the designed system, this paper proposes the use of regression methods for driver head pose estimation problem based on images from a stereo camera. Therefore, the system (Fig. 1) operates solely on the 3D head information and is independent of facial feature detection, color and texture information. The proposed regression algorithms for this work are Support

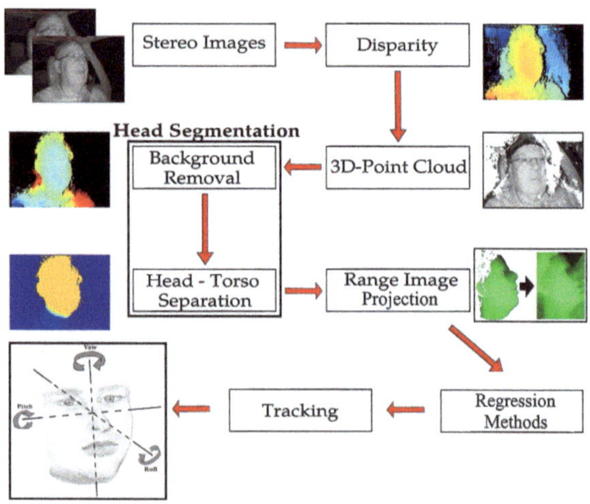

Fig. 1 Overview of the system's processing chain

Vector Regression (SVR), Random Regression Forest (RRF) and Extremely Randomized Trees (ERT). This paper proposes a reliable head pose estimation system, which can deal with obstacles that can be encountered in an automotive context. The experimental study demonstrates the high performance of the proposed algorithms which have outperformed the method employed in Höffken et al. [12] i.e. Synchronized Submanifold Embedding (SSE). The measured time-costs show the real-time capability of all of the methods. Moreover, the algorithmic simplicity of the proposed algorithms implies their suitability for embedded systems.

2 Related Work

Over the past few decades, there have been quite a number of studies which approached the head pose estimation problem from different perspectives. Most of the experimental approaches available in literature are carried out under the premise of a centrally aligned camera with respect to the subject in addition to ideal lighting conditions.

Oh and Kwak [18] use a centrally-aligned monocular infrared camera to capture the driver's image and implements Viola-Jones algorithms to generate classifiers that can detect the face region from the images. For facial features extraction, it employs linear discriminant analysis (LDAr) and a regression is done on this data set to obtain the estimated head poses. However, the system can still suffer from severe illumination changes and hard shadows. Murphy-Chutorian and Trivedi [17] combine a static head pose estimator and a head tracker. Based on an initial estimation of the head orientation, the system creates a texture-mapped 3D model of the head which is then rotated, translated and rendered to match each incoming new frames. Obviously, the system is computationally expensive and, in addition, the system can not tolerate large head pose variations.

In order to cope with large head turns and occlusions of facial features, the authors of [21] suggest the use of multiple cameras from different perspectives, while the head pose estimation is based on facial features and their respective geometry. Although multiple cameras might improve robustness with respect to large pose variations, they call for higher computational and financial costs.

Jiménez et al. [13] approach the head pose estimation problem by forming a sparse 3D face model acquired from the facial features. Nevertheless, this method also depends on the discernibility of the facial features. Krotosky et al. [14] propose the use of simple ellipse detectors to detect the driver's head from the disparity map formed from stereo camera images. The latter work confirms the applicability of disparity maps to deal with illumination changes and occlusions to a certain extent.

Subsequent to the introduction of 3D acquisition systems, it was possible to overcome the effects of illumination changes, shadows and occlusions using the available range images. Fanelli et al. [2] propose a feature-less approach that can estimate the pose parameters from all generic surface patches. Accordingly, [2] employed random regression forests for real time 3D head pose estimation on account of their ability to handle large training data with speed and accuracy. However, since a synthetic data set is used for the training, it lacks real time examples.

Fanelli et al. [3] extend the previous work and introduce an improvement by using classification to discriminate the depth image patches belonging to the head region and regression to predict the pose. This has resulted in better suitability to low quality depth data.

In Höffken et al. [12], nearest neighbor search in the feature space is performed for head pose estimation. Despite the simplicity of the algorithm and promising performance, nearest neighbor search suffers from the need for intensive memory access. Moreover, the method can not be employed with high dimensional feature inputs.

To our knowledge, there is no previous work employing stereo camera-based head pose estimation using extremely randomized trees.

3 System Overview

The designed system should consider real-time driving environments in order to be robust in practice. One of the many aspects is the position of the camera, which is not centrally aligned with respect to the driver inside the car due to design constraints and safety reasons. As depicted in Fig. 2b the stereo camera is mounted to the left of the navigation display on the dashboard of the car. Consequently, depending on the seat arrangement, the camera is located at 29°–40° with respect to the driver. Considering the size range of possible drivers, the vertical opening of the camera is 40°.

Furthermore, the designed system should also contemplate aspects such as: applicability on low-cost cameras with low resolution images, robustness against variations in illumination, shadow and occlusions, susceptibility to individual differences, no requirement for initialization and low computational demand for use in embedded systems.

To realize the system for the intended purpose, the input data passes through a series of preceding processes to form an input with the required format and information that can maximize the performance of the algorithms (Fig. 1).

3.1 3D Points Acquisition

The right and left images from the stereo camera show the driver from slightly different perspectives. Using the geometrical difference between the two pairs of gray-scale images, the 3D structure of the scene can be recovered.

The first step in the process is image rectification [4, 15], which is then followed by the computation of the disparity employing the FPGA based *Semiglobal*

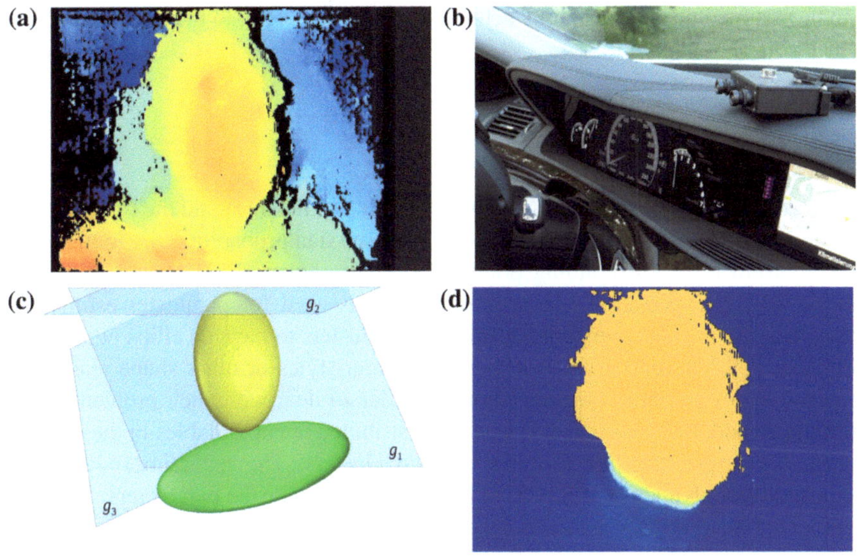

Fig. 2 Figure **a** shows the disparity image obtained from a pair of gray scale images from the stereo camera. **b** Demonstrates the position of the camera inside the car. The placement of the 3D planes that are used to remove the background with respect to the driver is given in (**c**). **d** Shows the final outcome of the head-torso separation

Matching approach [6, 10, 11]. Then, the 3D points can be accumulated in a point cloud structure. As can be seen in Fig. 2a, the disparity image disregards external elements such as changes in illumination and shadow, which is contributes to the robustness of the system.

3.2 *Head Segmentation*

This process involves the removal of the 3D data parts that do not provide information on the head of the driver. Therefore, this action comprises two steps: background removal comes first and head-torso separation follows leaving the head of the driver isolated.

3.2.1 Background Removal

Taking advantage of the vehicle's interior arrangements and the restricted movement of the driver, background subtraction can be achieved using three planes, which can be placed on the left, above and behind the driver's seat as shown in Fig. 2c. The two Planes (g_1, g_2) hold a fixed position while the third plane (g_3) has one degree of freedom depending on the individual seat adjustment. Therefore, it can be assumed that all the points behind these planes belong to the background and hence can be removed.

3.2.2 Head-Torso Separation

For head-torso separation, the $3D$ points belonging to the head and the torso regions are represented using a statistical model. Two Gaussian probability distributions are formed from the corresponding points of the two regions i.e. the head and torso points, where Expectation Maximization (EM) is used for parameter estimation. The Gaussian mixture components assume the clusters to have an ellipsoidal shape, therefore if the assumption is not met due to hair-styles or other shape reasons, it will result in unintended segmentation. In order to deal with such problems, [12] introduced additional normal and wish art distributed latent variables in the models, which are responsible for forming the two clusters. The main improvement is achieved through the use of Bayesian approach to estimate the parameters. The final segmented head is shown in Fig. 2d.

3.2.3 Dimensionality Reduction

The resulting range data from the solely head information is very high dimensional and thus unsuitable for most head pose estimation algorithms. Therefore, an additional procedure is implemented for dimensionality reduction and optimization of the data for further use.

To reduce the dimensions, a projection matrix is created which minimizes the distance between samples of similar head poses and maximizes the distance for samples of different head poses for the same individual. This is equivalent to Multiclass Linear Discriminant Analysis (M-LDA), where different head poses correspond to different classes. Finally, the obtained projection matrix will map the training and testing range image data into a lower-dimensional space.

The reduced feature dimension covers more than 98 % of head pose variances and is ordered according to the head pose related information. The objective behind this process is to reduce the dimensionality of the data and optimize the data for head-pose separation. Therefore, the optimized low dimensional data is used as an input for the regression algorithms that are used for the head pose estimation.

4 Regression Algorithms

To improve the robustness of the system, the head pose estimation problem is approached using regression methods. This means that for given training data $(x_1, y_1), \ldots, (x_n, y_n)$, where $x_i \in \mathbb{R}^d$ being the independent variable in terms of a range image and $y_i \in \mathbb{R}^k$ being the dependent variable in terms of a pose vector, what needs to be estimated is $f(x) = E(Y|X = x)$, which is the regression function of Y on X. The following subsections summarize three different powerful approaches to this regression problem.

4.1 Support Vector Regression

Support Vector Machines [20] have emerged as an effective technique for most regression problems. The most attractive characteristic of Support Vector Machines is that it employs the Structural Risk Minimization (SRM) principle, which maximizes the generalization ability of the algorithm [8]. Consider a given set of training data, $D = \{(x_1, y_1), \ldots, (x_l, y_l)\}$, where X is $\in R^n$ and $Y \in R$. In ε-insensitive regression, the objective is to find a function that is as flat as possible and approximates the input patterns x_i from the labeled targets y_i with at most ε deviation. This means any error within the ε bound is tolerated.

The desired approximation function is given by $f(x, \alpha) = w \cdot \phi(x) + b$, where ϕ is a function that introduces nonlinearity by a map from the input space into a high dimensional feature space, α denotes a set of parameters and b is a constant parameter [19]. As the linear function is dependent only on the dot product of the input patterns x_i, it is sufficient to know $\langle \Phi(x), \Phi(x') \rangle$, which is equivalent to evaluation of kernel functions, rather than the computationally demanding explicit mapping. Hence, by using kernel function i.e. $k(x, x')$, the optimization problem can be given as:

$$\underset{\alpha,\alpha^{(*)}}{\text{maximize}} \begin{cases} -\frac{1}{2}\sum_{i\,j=1}^{l}(\alpha_i - \alpha_i^*)(\alpha_j - \alpha_j^*)k(x_i, x_j) \\ -\varepsilon\sum_{i=1}^{l}(\alpha_i + \alpha_i^*) + \sum_{i=1}^{l} y_i(\alpha_i - \alpha_i^*) \end{cases} \tag{1}$$
$$\text{subjected to} \quad \sum_{i=1}^{l}(\alpha_i - \alpha_i^*) = 0 \text{ and } \alpha, \alpha^* \in [0, C]$$

The prediction function resulting from the optimization problem is given by:

$$w = \sum_{i=1}^{l}(\alpha_i - \alpha_i^*)\Phi(x_i) \quad \text{and} \quad f(x) = \sum_{i=1}^{l}(\alpha_i - \alpha_i^*)k(x_i, x_j) + b \tag{2}$$

Equation (2) implies, the solution is solely dependent on the dot products of the input values with non-zero $\alpha, \alpha^{(*)}$ coefficients. These points are the training patterns lying on the boundary or outside the ε tube since the points inside the tube have $\alpha, \alpha^{(*)} = 0$. Therefore, only those points, the so called *Support Vectors*, are involved in the prediction. Accordingly, the complexity of the model is not dependent on the dimensionality of the input data but rather on the number of support vectors [20].

4.2 *Random Regression Forest*

Tree-based ensemble methods [1] are a collection of trees in which the final prediction is an aggregation of the decisions of the individual trees. Random forests have added extra randomness that has a significant influence on the accuracy of the estimation and is achieved through two procedures [9]. Primarily, the fitting of the trees is done using bootstrap samples $D_1, \ldots, D_B$ of size N' from the original dataset D of size N. Furthermore, additional randomness is added during the splitting of the nodes. Because of their low computational costs and high accuracy for large data sets, random forest has appeared to be a commonly employed method for regression problems in recent researches [3]. The pseudo code given in (1) illustrates the working flow of the algorithm.

Algorithm 1 Random Forest for regression

1. During training For $b = 1$ to B trees:
   a. Draw a bootstrap sample $\mathbb{Z}^*$ of size N from the original data.
   b. Grow the individual trees T_b in the forest from the bootstrapped data, by recursively repeating the following step at each node of the tree, until the minimum sample count (n_{min}) is reached.
      i. Select m variables randomly from the total p predictor variables.
      ii. Pick the variable that optimizes the split among m.
      iii. Split the node into two sub-nodes.
2. For prediction on a new sample, direct the sample towards the leaf according to the test at each node of the trees. Output the ensemble of the leaf nodes the sample has reached $\{T_b\}_1^B$.

For regression, the prediction of the leaf nodes at a new point x is given by:

$$f(x) = \frac{1}{B} \sum_{b=1}^{B} T_b(x) \tag{3}$$

4.3 *Extremely Randomized Trees*

Extremely Randomized Trees were first introduced in [7]. They are a collection of unpruned trees, where the whole data set is used for the fitting of the trees in contrast to random forests where bootstrap samples are used to build the trees. The pseudo code given in (2) clarifies the working principle of the algorithm [7].

Algorithm 2 Extremely Randomized Trees for regression

1. For $b = 1$ to B trees:
   a. Grow the individual trees T_b from the training data N, by recursively repeating the following step at each node of the tree, until the minimum sample count (n_{min}) is reached.
      i. Select randomly m variables, $\{a_1, ..., a_m\}$, without replacement, from the total p predictor variables.
      ii. For each selected variable a_i $\forall i = 1, ..., m$:
         A. Compute the maximum and minimum value of a_i denoted as a^N_{max} and a^N_{min}.
         B. Draw a random cut-point s_i uniformly in $[a^N_{min}, a^N_{max}]$.
         C. Return the split $[a_i < s_i]$.
      iii. Among the m splits $\{s_1, ..., s_m\}$, select the split s_* such that

$$s_* = \underset{i=1,...,m}{\arg\max} \quad \Delta_{var}(s_i, N) \tag{4}$$

2. For prediction on a new sample, direct the sample towards the leaf according to the test at each node of the trees. Output the ensemble of the leaf nodes the sample has reached $\{T_b\}_1^B$.

For regression, the prediction of the leaf nodes at a new point x is given similarly by:

$$f(x) = \frac{1}{B} \sum_{b=1}^{B} T_b(x) \tag{5}$$

As described in (2), the splitting of the nodes is done in such a way that for each randomly selected variable subset m from the total subset, a random splitting value (cut-point) is chosen. Among the available m splits, the split that maximizes the variance reduction [5] given by

$$\Delta_{\text{var}}(s_i, N) = \{y|N\} - \frac{|N_l(a_i)|}{|N|} \text{var}\{y|N_l(a_i)\} - \frac{|N_r(a_i)|}{|N|} \text{var}\{y|N_r(a_i)\} \tag{6}$$

is selected, where $N_l(a_i)$ and $N_r(a_i)$ are the two subsets of N satisfying $a_i < s_i$ and $a_i \geq s_i$, and $|N|$ the total number of training samples.

5 Experiments and Results

The total training dataset is comprised of a captured images of 39 individuals and was recorded at 25 fps using a stereo camera within the simulated driving environment. The driving scenario simulates countryside as well as highway traffic and also incorporates several distractive tasks such as paying attention to certain passing objects and operating the center stack display. All sequences have an approximate length of 1 h and contain subjects of varying age, gender and race. Two subjects are wearing glasses.

The ground truth information about each subject's head pose has been simultaneously recorded using a high precision laser tracker[1] and transformed to the car coordinate system. Quaternions are used to describe the head rotation, which is later used as a label space during the training.

The prediction model should learn the optimal parameters for the best possible performance when predicting on new data. A *3-fold cross validation* is used to choose the optimal parameters from the desired ranges of parameters for each algorithm. The parameters that yield the minimum *root-mean-square-error (RMSE)* are selected as optimal parameters.

For SVR, linear and Radial Basis Function (RBF) are selected from the commonly used kernel functions to be examined for this paper. A linear kernel was used on the data to benefit from lower computational cost if the data is linearly regressible. However, there is no guarantee that a linear kernel achieves the maximum possible performance gain, so using an RBF kernel can introduce nonlinearity. RBF kernel was selected because it is the most recommended nonlinear kernel for similar problems. The degree of the RBF kernel is set to 2. The optimized parameters for SVR are: the ε deviation, the penalty cost c and the γ parameter for the RBF kernel. The optimal parameters are found to be $\varepsilon = 0.0082$, $c = 0.0151$ and $\gamma = 0.011$.

In the experiment, the number of trees for RRF is set to 100 to make sure it is above the threshold value, the point where the error converges. Additionally, the maximum depth the trees can grow to is set to 20. The optimized parameters are the number of predictor variables m and minimum sample count $n_{\min}$ and are found to be 5 and 17 respectively. The optimal value for the number of predictor variables shows that for the suggested value for regression [9], one-third of the total variables i.e. 3, is sub-optimal for the problem at hand.

For ERT, in a similar way as RRF, the number of trees was set to 100 and maximum depth to 20. The obtained optimal parameters are: number of predictor variables $m = 10$ and minimum sample count $n_{\min} = 7$.

After the optimal parameters are found, the performance of the algorithms is evaluated using a *leave-one-individual-out cross validation*. Leave-one-individual-out is employed in such a way that the data from the 38 individuals are put together and used as a training set, whereas the data from one individual is a test set.

Table 1 compares performances and time-costs of all proposed algorithms including SSE. It can be observed that SVR (RBF kernel) has performed significantly better than SVR (Linear kernel). Evidently, the different regression algorithms that are studied for the purpose have more or less comparable prediction accuracy. However, as given in Table 1, the algorithms have outperformed the implemented method in [12] i.e. SSE. In addition, Fig. 3 visualizes the projected absolute angular pose estimation error in relation to varying yaw and pitch displacements over the ground truth data. The accuracy of the algorithms for most of the samples with less pitch and yaw angular movements is much better than for

[1]LaserBIRD, Ascension Technology Corporation.

Table 1 The mean absolute angular error and a comparison in prediction accuracy of SVR (linear and RBF kernel), RRF, ERT and SSE

Algorithms	Mean error in degree	Accuracy in %			Time-cost in ms/frame
		≤15°	≤10°	≤5°	
SVR—linear	8.42	90.64	78.60	34.94	1.1
SVR—RBF	6.96	94.38	85.34	45.42	4.7
RRF	6.82	94.93	86.17	46.23	1.2
ERT	6.79	95.03	86.33	46.67	1.3
SSE	7.41	93.91	83.12	40.02	1.5

Prediction errors less than the given angle thresholds are considered to be correctly estimated

samples with large head movements. Nevertheless, inferring from the ground truth, most of the samples have a yaw movement between −35° and +35° and a pitch movement of −5° to +30° implying lack of training data for larger head movements, which explains the relatively worse performances of the algorithms for larger head movements. Figure 4 compares the predictions of all algorithms for a random subset of individuals. The figure shows that for each individual the regression methods generally perform better than SSE.

The computation time of the algorithms is highly dependent on the chosen model parameters. For SVR, a linear kernel might be a safe choice regarding model complexity and time-cost, however it resulted in a relatively worse prediction accuracy. In contrast, when the RBF kernel is employed, the model complexity

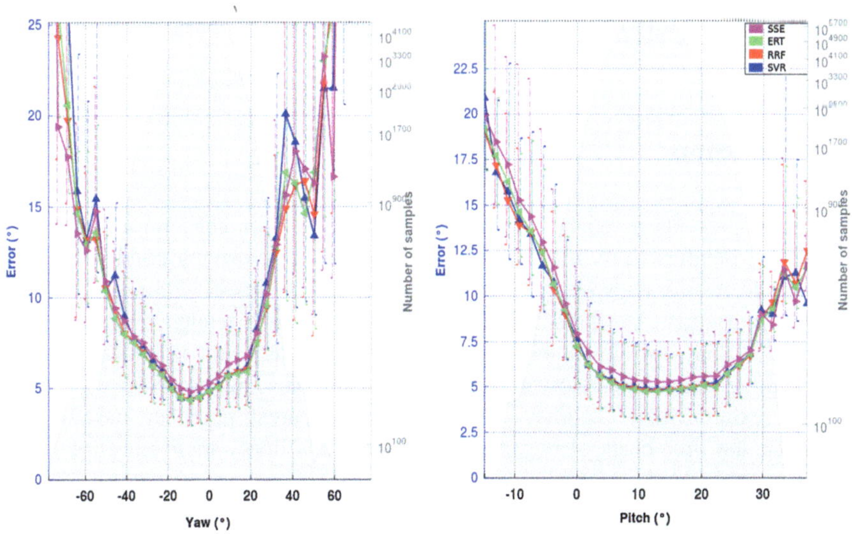

Fig. 3 A comparison among the projected error of support vector regression (RBF kernel), random regression forests, extremely randomized trees and synchronized submanifold embedding over the ground truth data given in yaw and pitch angle

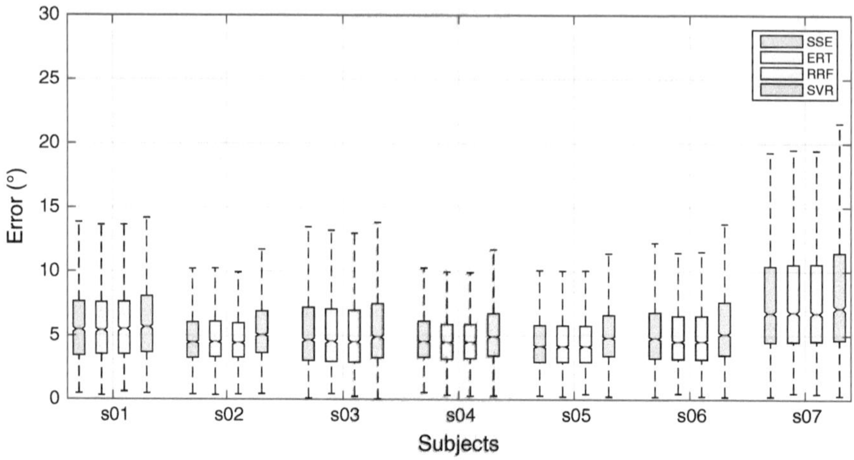

Fig. 4 A box plot of performance comparison for a subset of individuals with respect to the absolute angular error

increases quadratically, but an improvement in accuracy is achieved. The same argument applies for the tree-based ensemble methods. During the experiment, it was possible to achieve a similar performance using only 10 trees proving that an increase in trees is only essential to a certain extent, thereafter it is only increases time cost.

The time-cost measurements given in Table 1 were conducted on a single core *Intel i7* with 3.20 GHz. They indicate the real-time capability of all the presented methods also on embedded systems.

6 Conclusion

This paper presents a reliable stereo camera based driver head pose estimation system that handles many difficulties encountered in real driving conditions. The training and testing data were acquired using a stereo camera within a simulated driving environment in which 39 individuals with different genders and sizes took part. The system operates solely on 3D head information and is independent of facial feature detection, color and texture information. The proposed regression methods exhibit high accuracy and outperform nearest neighbor based SSE. Moreover, the time-costs of the employed algorithms imply their real-time capabilities on embedded systems.

Based on the enhanced frame-wise head pose estimation approaches presented here, further improvements in accuracy and robustness will be achieved by applying a quaternion motion tracker.

References

1. Cutler A, Cutler DR, Stevens JR (2009) Tree-based methods. In: High-dimensional data analysis in cancer research. Springer, Berlin, pp 1–19
2. Fanelli G, Gall J, Van Gool L (2011) Real time head pose estimation with random regression forests. In: 2011 IEEE conference on computer vision and pattern recognition (CVPR). IEEE, pp 617–624
3. Fanelli G, Weise T, Gall J, Van Gool L (2011) Real time head pose estimation from consumer depth cameras. In: Pattern recognition. Springer, Berlin, pp 101–110
4. Fusiello A, Trucco E, Verri A (2000) A compact algorithm for rectification of stereo pairs. Mach Vis Appl 12(1):16–22
5. Galelli S, Castelletti A (2013) Assessing the predictive capability of randomized tree-based ensembles in streamflow modelling. Hydrol Earth Syst Sci Dis 10(2):1617–1655
6. Gehrig SK, Eberli F, Meyer T (2009) A real-time low-power stereo vision engine using semi-global matching. In: Computer vision systems. Springer, Berlin, pp 134–143
7. Geurts P, Ernst D, Wehenkel L (2006) Extremely randomized trees. Mach Learn 63(1):3–42
8. Gunn SR et al (1998) Support vector machines for classification and regression. ISIS technical report, 14
9. Hastie T, Tibshirani R, Friedman J, Friedman J, Tibshirani R (2009) The elements of statistical learning, vol 2. Springer, Berlin
10. Hirschmuller H (2005) Accurate and efficient stereo processing by semi-global matching and mutual information. In: IEEE computer society conference on computer vision and pattern recognition, 2005. CVPR 2005, vol 2. IEEE, pp 807–814
11. Hirschmuller H (2008) Stereo processing by semiglobal matching and mutual information. IEEE Trans Pattern Anal Mach Intell 30(2):328–341
12. Höffken M, Tarayan E, Kresel U, Dietmayer K (2014) Stereo vision-based driver head pose estimation. In: 2014 IEEE Intelligent Vehicles Symposium Proceedings. IEEE, pp 253–260
13. Jiménez P, Bergasa LM, Nuevo J, Hernandez N, Daza IG (2012) Gaze fixation system for the evaluation of driver distractions induced by ivis. Intell Transp Syst IEEE Trans 13(3):1167–1178
14. Krotosky SJ, Cheng SY, Trivedi MM (2005) Real-time stereo-based head detection using size, shape and disparity constraints. In: Intelligent Vehicles Symposium, 2005. Proceedings. IEEE. IEEE, pp 550–556
15. Krueger L (2007) Model based object classification and localisation in multiocular images. 2301391
16. Murphy-Chutorian E, Doshi A, Trivedi MM (2007) Head pose estimation for driver assistance systems: a robust algorithm and experimental evaluation. In: IEEE intelligent transportation systems conference, 2007. ITSC 2007. IEEE, pp 709–714
17. Murphy-Chutorian E, Trivedi MM (2008) Hyhope: hybrid head orientation and position estimation for vision-based driver head tracking. In: 2008 IEEE intelligent vehicles symposium. IEEE, pp 512–517
18. Oh JH, Kwak N (2012) Recognition of a driver's gaze for vehicle headlamp control. Veh Technol IEEE Trans 61(5):2008–2017
19. Rajwade A, Levine MD (2006) Facial pose from 3D data. Image Vis Comput 24(8):849–856
20. Smola AJ, Schölkopf B (2004) A tutorial on support vector regression. Stat Comput 14 (3):199–222
21. Tawari A, Martin S, Trivedi MM (2013) Monitoring head dynamics for driver assistance systems: a multi-perspective approach. In: International Annual Conference on Intelligent Transportation, 2013 IEEE. IEEE

Future Computer Vision Algorithms for Traffic Sign Recognition Systems

Stefan Eickeler, Matias Valdenegro, Thomas Werner and Michael Kieninger

Abstract For the assistance of drivers and for autonomous vehicles an automatic recognition of traffic signs is essential. Today's traffic sign recognition systems are focusing on circular signs. Because speed limit signs are circular and because the recognition of speed limit signs deliver a high customer benefit, such systems have been realized primarily in series production. But for traffic sign recognition there are still many challenges to be tackled. This contribution presents how research and intelligent development of computer vision algorithms will enable much more advanced traffic sign recognition systems on embedded systems by reducing processing time while simultaneously enlarging functionality. The result is a reduction of hardware cost and energy consumption.

Keywords Symmetry detector · Traffic sign detection · Traffic sign recognition

1 Introduction

The process of traffic sign recognition can be divided into two steps: detection and recognition (classification). At actual systems the detection step needs a lot of computing power. For speed limits the detection of circular signs is necessary.

S. Eickeler (✉) · T. Werner · M. Kieninger
Fraunhofer Institute for Intelligent Analysis and Information Systems IAIS, Schloss Birlinghoven, 53754 Sankt Augustin, Germany
e-mail: stefan.eickeler@iais.fraunhofer.de

T. Werner
e-mail: thomas.werner@iais.fraunhofer.de

M. Kieninger
e-mail: michael.kieninger@iais.fraunhofer.de

M. Valdenegro
Hochschule Bonn-Rhein-Sieg, Grantham-Allee 20, 53757 Sankt Augustin, Germany
e-mail: matias.valdenegro@gmail.com

T. Schulze et al. (eds.), *Advanced Microsystems for Automotive Applications 2015*, Lecture Notes in Mobility, DOI 10.1007/978-3-319-20855-8_6

The classical method for circle detection is the use of the circular Hough transformation. In [1] a fast version of this transformation was presented as Fast Radial Symmetry Detection, representing the current state of the art. In order to achieve a real-time detection, the algorithm is processed on parallel processors or on dedicated hardware.

In this paper firstly we present how this approach of circle detection is significantly improved in order to minimize processing time, memory and energy consumption by our methods consisting of complexity reduction by new methods, computing time reduction by heuristics, runtime optimization on the target system and developing of portable C/C++ algorithms. These achievements contribute to solve the situation, that algorithms from current research require many resources e.g. 10 times more than available. And according to Moore's Law 10 times performance would lead to 5 years waiting time before the developed technology can be introduced to market.

Secondly we present results using deep neural networks for recognition of traffic signs. We found many indicators that this classifier delivers much better results than conventional classifier regarding precision, recall, model size and computing time.

Thirdly we present how the modification of our circle detector will lead to a high performance detector for arbitrary regular polygons. Regarding autonomous driving it will be indispensable to detect and recognize all types of road signs. Especially regarding more complex signs not only the recognition but also the interpretation of signs containing text or detour recommendations will be indispensable. To tackle this challenge we show approaches using optical character recognition methods.

2 Radial Symmetry Detector

The Radial Symmetry Detector (RSD) is a gradient-based, Hough-like algorithm for the detection of circles of a range $r \in R$ of known radii. The algorithm [1–3] works as follows: (i) First, the input image is smoothed using a Gaussian filter with a small filter size. (ii) Second, using the Sobel operator the gradient magnitude and orientation of each pixel is computed. (iii) Next, based on the gradient information, a voting mechanism is used to estimate the circle centers based on the different possible radii.

The voting process constructs a 2D accumulator array O_r of the size of the input image for every possible radius and performs voting as follows: Each pixel p with a gradient magnitude larger than a threshold G_t is assumed to be part of a circle and votes for a corresponding circle center p_+ based on the range of radii. A vote for a center location is an increment of the corresponding cell in the radius-specific accumulator array.

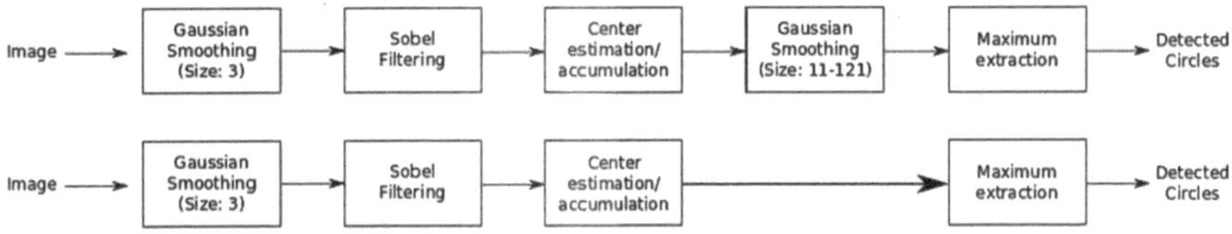

Fig. 1 Overview of the processing pipeline of the original and the improved radial symmetry detector

$$p_+ = p + round\left(r \frac{g(p)}{||g(p)||}\right)$$

$$p_- = p - round\left(r \frac{g(p)}{||g(p)||}\right)$$

To detect dark circles on light background and vice versa, an additional negative vote p_- is counted on the opposite side of the circle.

$$O_r(p_+) = O_r(p_+) + 1$$

$$O_r(p_-) = O_r(p_-) - 1$$

After the accumulator arrays are computed, the symmetry contribution images F_r are derived and smoothed using another Gaussian filter A_r:

$$F_r = \left(\frac{O_r}{k_r}\right)^{\alpha} * A_r.$$

Here, k_r is a size dependent normalization constant and α the radial strictness parameter with a value of 2 [1]. The Gaussian filter is chosen to have a size of $r \times r$ and a standard deviation of $\sigma = 0.5r$. Finally, the absolute maxima in the smoothed symmetry contribution images are extracted, which represent the centers of circles present in the input image. An overview of the processing pipeline of the RSD can be seen in Fig. 1.

2.1 *Improved Radial Circle Detector*

Since the final smoothing of the symmetry contribution image (step iv) is the most computational expensive with about 90 % of the processing time, it has the highest potential to benefit from optimization. The optimized method replaces the fixed size accumulator arrays, with version scaled by $\frac{r_{base}}{r}$, where r_{base} is the smallest radius that should be detected. As a results the memory consumption of the accumulator is almost halved. Since the size of the accumulator changed, a modified voting scheme must be applied. Instead of changing a single accumulator cell, 2×2 cells are changed at once by not only (de)incrementing the candidate center cell but also its direct neighbours by

$$O_r(p_+) = O_r(p_+) + 1 \quad \text{and} \quad O_r(p_+ + v) = O_r(p_+ + v) + 1$$
$$O_r(p_-) = O_r(p_-) + 1 \quad \text{and} \quad O_r(p_- + v) = O_r(p_- + v) - 1$$

with $v \in \left\{ \begin{pmatrix} 0 \\ 0 \end{pmatrix}, \begin{pmatrix} 1 \\ 0 \end{pmatrix}, \begin{pmatrix} 0 \\ 1 \end{pmatrix}, \begin{pmatrix} 1 \\ 1 \end{pmatrix} \right\}$.

Since the accumulator arrays are scaled according to the detection radius, the scaling has also to be incorporated into the computation of the possible circle centers.

$$p_{\pm} = round\left(r_{base} \left(\frac{1}{r} p \pm \frac{g(p)}{\|g(p)\|} \right) \right)$$

The usage of smooth 2×2 updates in the scaled accumulator arrays makes it possible leave out processing (step iv) from the original pipeline, because it implicitly resembles the Gaussian smoothing operation under the assumption of isotropic errors [4]. Another benefit of the scales accumulators is that subsequent operation will perform fast due to the reduced size.

An additional simplification is applied during the computation of F_r by using values of $k_r = r$ and $\alpha = 1$. As a consequence the computational effort is reduced.

To make the extraction of the circle centers more robust, the accumulators for three consecutive radii are combined and the absolute maxima are extracted from there. This has the additional effect, that circle radii that do not match the tested radii can also be detected reliably.

In Fig. 2 the accumulator arrays for a test image can be seen. The detection of the two circular structures is mainly based on the last three scales.

3 Improved Regular Polygon Detector

The Regular Polygon Detector (RPD) is very similar to the RSD, but has two main differences. The first difference is that r is the radius of the apothem of the polygon. An apothem is a circle that is tangent to every side of the polygon. This essentially means that the RSD scheme is used to detect the apothem of a polygon instead of circles. The second difference is the counting scheme used to fill the accumulator. Instead of voting for a single center location, the cells of several possible locations along a vector perpendicular to the gradient are modified. The voting line is defined as

Fig. 2 Example accumulator arrays for a real life scene. It can be seen that the *blue circular sign* as well as the circular structure within the *white sign* are detected, especially in the last three scales

$$L(p,m) = p \pm round\left(r\frac{g(p)}{||g(p)||}\right) \pm round(mg_{\perp}(p))$$

Here, $g_{\perp}(p)$ is a unit vector perpendicular to g(p). The width w of the voting line is given by

$$w = r * \tan\left(\frac{\pi}{n}\right),$$

where n is the number of sides of the regular polygon being detected. Pixels on the line where $m \in [-w, w]$ receive a positive vote and pixel with $m \in [-2w, -w-1]$ or $m \in [w+1, 2w]$ will receive a negative vote. After the voting process, images similar to the symmetry contribution images are computed and used to extract the centers of the apothems. In contrast to the RSD, no additional Gaussian smoothing is applied due to the high computational effort.

Since RPD and RSD use the same algorithmic principles, the same optimization steps as for the RSP can be applied here. The accumulator arrays are again scaled with respect to the size of the smallest object that should be detected, and a modified voting scheme, that incorporates that, is applied. The new voting line given by

$$L_{\pm}(p,m) = round\left(r_{base}\left(\frac{p}{r} \pm \frac{g(p)}{||g(p)||} \pm mg_{\perp}(p)\right)\right).$$

The length of the voting line also has to be modified accordingly and is defined as

$$w = \frac{r_{base}}{r}\tan\left(\frac{\pi}{n}\right).$$

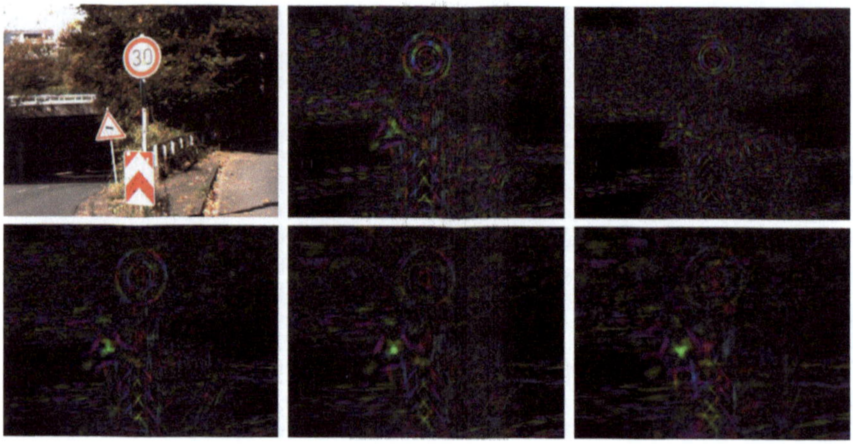

Fig. 3 Example accumulator arrays for a given test image. It can be seen that especially in the last three scales the *triangular signs* are nicely detected

Example accumulator arrays for a test image are shown in Fig. 3. It can be seen that the triangular signs are nicely detected.

4 Traffic Sign Recognition Using Deep Neural Networks

The RSD can be extended by a recognition stages to recognize round traffic signs. A method to realize this is convolutional neural networks (CNN), based on Hubel and Wiesel's findings of receptive field in the visual cortex. CNNs have shown to be very powerful in solving general visual recognition tasks and traffic sign recognition in particular [6]. The proposed system uses a simple deep network structure of convolutional and pooling layers and is trained to accept round traffic singes while rejecting other circular objects. An overview of the structure of the net can be seen in Fig. 4.

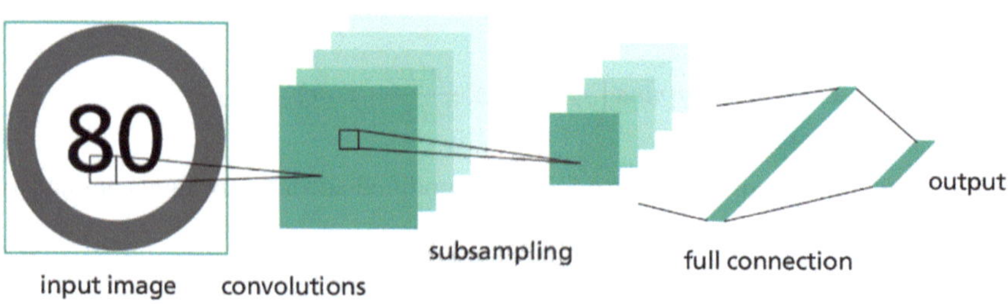

Fig. 4 Network structure used for the recognition of traffic signs

Table 1 Comparison of processing speed of different RSD implementations

Paper	Implementation	Resolution	Frames/s	Pixels/s (M)
[3]	MMX, multi-processing	320 × 240	30	2.3
[5]	GPU	640 × 480	33	10.1
Ours	C++ single threaded	800 × 600	54	25.9

5 Evaluation

This section gives an overview of the results of the proposed methods. First, the improved and original RSD and RPD are compared in terms of runtime and accuracy on synthetic images, and finally the recognition performance on the German Traffic Sign Benchmark [8] is evaluated.

5.1 RSD and RPD

The proposed optimized RSD is compared to the original algorithm and an additional GPU-based implementation. In Table 1 it can be seen that the original algorithm has a relatively slow processing speed of 2.3 M pixel/s whereas the optimized version is more than 10 times as fast with 25.9 M pixel/s. Although for the evaluation of the optimized version an out-dated CPU was used, a precise comparison is not possible. To accomplish this, the original version was re-implemented in C++, programmatically optimized like the optimized version and compared to it using several artificial test images containing a single circle and additive Gaussian noise with specific deviations.

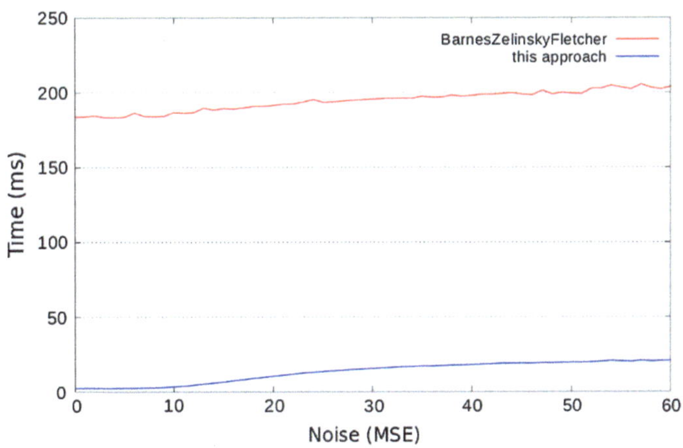

Fig. 5 Comparison of the processing time of the original and the improved RSD with respect to the added noise

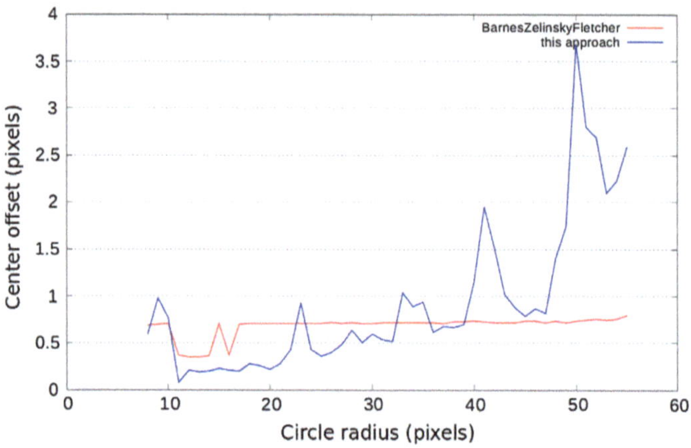

Fig. 6 Comparison of the detection accuracy (in pixels) of the original and improved RSD with respect to the circle radius

In Fig. 5 the runtime of both implementations is compared with respect to the additive noise. It can be seen that when the noise is below the gradient threshold, the computation time of the original method is dominated by the Gaussian smoothing step. Since this processing step is missing in the optimized version, it performs the detection a lot faster. As soon as the noise exceeds the threshold, the processing time of both version increases.

A comparison on real world images shows a speed up factor of 13 but the same detection rate of the optimized version compared to the original.

Another criterion that was tested, was the detection accuracy (offset detected center to actual center). In Fig. 6 the accuracy is plotted against the circle radius. It can be seen that the original method has almost a constant accuracy, whereas the proposed system has constant accuracy regarding the circle radius.

5.2 *Recognition Using Deep Neural Networks*

The evaluation of the complete recognition system was performed on the German Traffic Sign Benchmark [7]. It consists of 300 annotated traffic scenes, recorded by a dashboard mounted camera and is split up into different sign categories. The proposed method achieves a recall of 99.4 % and a precision of 96.3 % on the Mandatory subset. The corresponding recognition rate is 97.5 %.

6 Conclusion

We have shown how the Radial Symmetry Detector can be improved to increase processing speed by a factor of 13. Although we observed a loss of detection accuracy on artificial images, this effect had no impact on traffic sign recognition systems using real life images.

Additionally we presented a similar optimization for the Regular Polygon Detector. To overcome false detections, a deep neural network was used to verify the detected circles as traffic signs. A combination of our improved Radial Symmetry Detector and a deep neural net achieved a recognition rate of 97.5 % on the German Traffic Sign Benchmark.

References

1. Loy G, Zelinsky A (2003) Fast radial symmetry for detecting points of interest. IEEE Trans Pattern Anal Mach Intell
2. Banes N, Zelinsky A (2004) Real-time radial symmetry for speed sign detection. In: Intelligent vehicle symposium
3. Banes N, Zelinsky A, Fletcher L (2008) Real-time speed sign detection using the radial symmetry detector. In: IEEE transactions on intelligent transportation systems
4. Ballard D (1981) Generalizing the hough transformation to detect arbitrary shapes. Pattern Recogn
5. Glavtchev V, Muyan-Ozcelik P, Ota J, Owens J (2011) Feature-based speed limit sign detection using a graphics processing unit. In: IEEE intelligent vehicles symposium
6. Sermanet P, LeCun Y (2011) Traffic sign recognition with multi-scale convolutional networks In: International joint conference neural networks
7. Houben S, Stallkamp J, Salmen J, Schlipsing M, Igel C (2013) Detection of traffic signs in real-world images: the German traffic sign detection benchmark. In: International joint conference on neural networks

Inertial Sensors Integration for Advanced Positioning Systems

Marco Ferraresi

Abstract Modern navigation systems for the automotive market do require accurate and reliable sensors for precise implementation of the dead-reckoning algorithms needed for accurate map positioning in all different driving situations and circumstances. STMicroelectronics has been developing inertial sensors for these applications since 2009, with the introduction of the first 3-axis low-g accelerometers and the first ever fully integrated 3-axis gyroscope, now accepted as market standards and adopted in a myriad of in-dash infotainment systems. As happened earlier in other segments such as consumer, sensor integration is emerging in automotive to optimize performance and costs. Complete integration of acceleration and rate sensors on the same die can be enabled only by an advanced planar silicon technology, which is capable of creating independent mechanical structures that are diverse by nature. The tiny ASM330LXH is the smallest and highest-performance 6-axis Inertial Measurement Unit qualified to AEC-Q100 for non-critical automotive applications like navigation and telematics. This paper addresses the major integration challenges and presents the new device, with focus on the benefits versus traditional discrete solutions.

Keywords Navigation · Dead reckoning · Gyroscope · Accelerometer · Pressure sensor · Integration · Getter · Inertial sensors

1 Inertial Sensors and Positioning Systems

Inertial sensors are at the core of today's advanced positioning systems. A degradation of the signal coming from satellites may lead to a progressive loss of accuracy, which eventually prevents the operability of the system.

M. Ferraresi (✉)
STMicroelectronics, Via Tolomeo 1, 20010 Cornaredo, MI, Italy
e-mail: marco.ferraresi@st.com

T. Schulze et al. (eds.), *Advanced Microsystems for Automotive Applications 2015*,
Lecture Notes in Mobility, DOI 10.1007/978-3-319-20855-8_7

Dead Reckoning (aka DR) algorithms [1], based on sophisticated mathematical models, are fueled from the information gathered from a set of sensors, namely distance sensors (odometer, wheel speed, ABS wheel tick) and yaw rate sensors (gyroscope, differential wheel pulses, differential wheel speed).

STMicroelectronics has been at the forefront of the developments in this area for several years, both with the launch of the Teseo™ satellite position receivers and with the introduction of the first multi-axial inertial sensors [2, 3].

ST turnkey solution based on TeseoII, equipped with a proprietary DR algorithm, with a 6-axis sensing capability given by A3G4250D (3-axis gyroscope) and AIS328DQ (3-axis low-g accelerometer), is a widely recognized platform in the market, used by key players in the industry.

ST's Multi Sensor Dead Reckoning (MSDR) generates accurate position, heading, height and speed measurements, providing a fusion of GNSS with a number of sensors, bringing strong value to customer experience (Fig. 1):

- Automatic Temperature Compensation (ATC): it compensates each sensor for thermal effects, guaranteeing long-term accuracy (up to 90 %) even in the absence of the GNSS signal
- Automatic Free Mount (AFM): the tilt angle of the PCB is measured by the 3-axis accelerometer and input in, thus guaranteeing performance independent of the way the device is mounted on the vehicle
- Sensor Over UART (SOA): this allows the customer's host processor to collect sensor data and feed them to Teseo DR SW via UART
- 3D DR (3DR): this provides an indication of position variation in vertical direction, even in the absence of the GNSS signal

The latest satellite positioning receiver from ST, the TeseoIII, is capable of receiving signals from all existing satellite navigation systems, including GPS (USA), Galileo (Europe), GLONASS (Russia), QZSS (Japan), and BEIDOU (China) [4, 5].

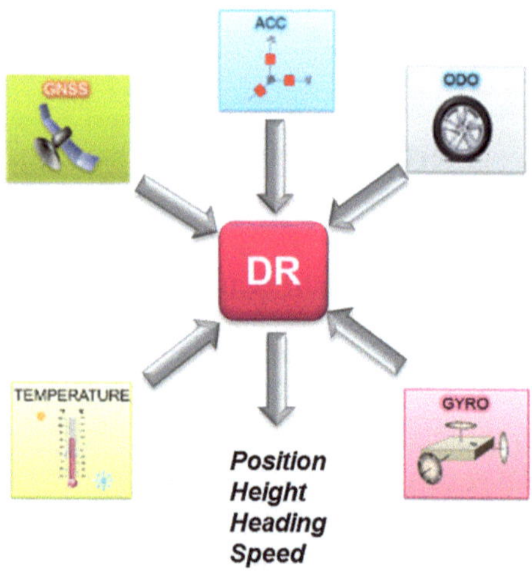

Fig. 1 Dead reckoning approach at ST

2 Sensors Integration: Key Challenges

Sensors integration has been applied in the consumer and mobile market for a few years already [6, 7]. Advanced smartphones, for instance, are equipped with multi-axial sensors with 6 or even 9 Degree of Freedom (DOF).

Automotive system makers, driven by the requests coming from car makers to reduce overall cost and improve sensing performance, are now pushing silicon manufactures to pursue an increase in component integration.

This is not an easy task to achieve; while it is almost straightforward to assemble multiple dies in a single and bigger package (Fig. 2), it is more challenging to reduce the number of dies and target a package with a small footprint (Fig. 3).

Silicon technology is the enabler of such a high level of integration.

ST's proprietary MEMS technology, called THELMA (Thick Epitaxial Layer for Micro-gyroscopes and Accelerometers), allows the integration, on the same piece of silicon, two diverse-by-nature mechanical structures like accelerometers and gyroscopes.

This technology guarantees an accurate control of all micro-mechanical features by using manufacturing and control techniques typical of IC processes.

Two silicon substrates are bonded together: the first one with the actual mechanical structure and the second one (the cap) to protect the movable elements during the over-molding process. The sensor micro-cavity is sealed and maintained throughout life at controlled atmosphere and pressure.

A getter layer (a patented alloy made of Zirconium and Vanadium) is included in the wafer-level hermetic package of gyroscopes to achieve a stable low bonding pressure value (<1 mbar) and therefore obtain low damping and reduced power consumptions for the MEMS resonators (Fig. 4).

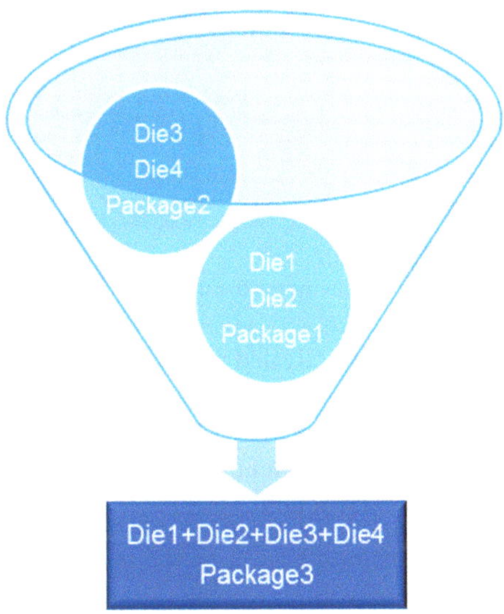

Fig. 2 Traditional SiP solution

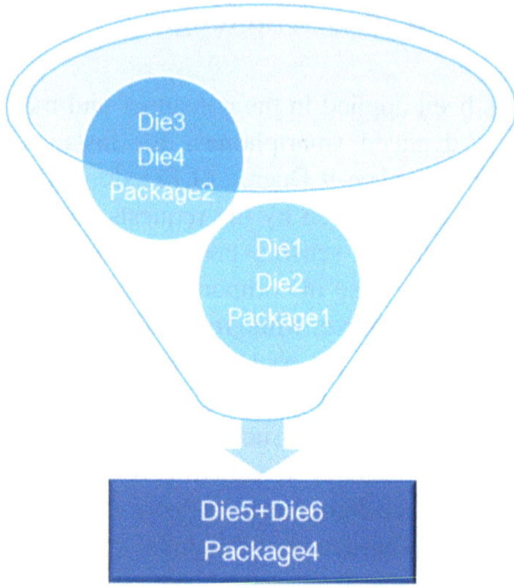

Fig. 3 Advanced system integration

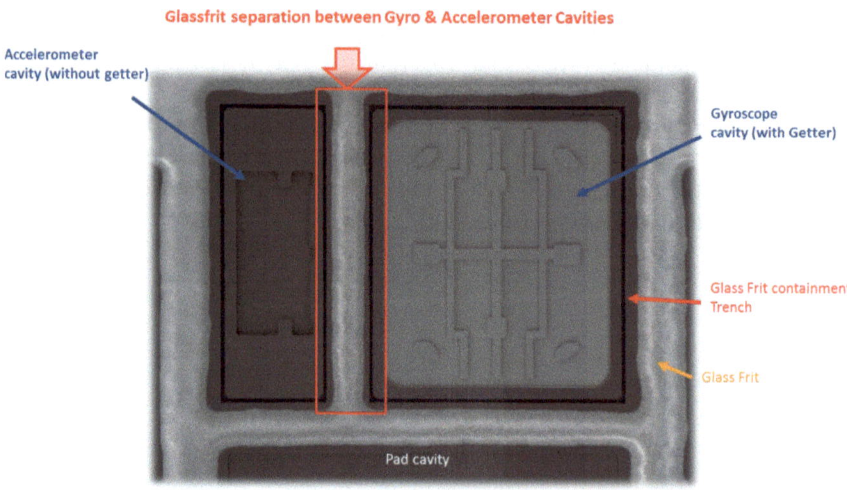

Fig. 4 Double cavity for 6-axis inertial module

After being activated with specific thermal treatments, the getter layer absorbs all active gases present in the hermetic cavity, such as O_2, CO, CO_2, N_2, leaving only noble gases in the cavity (Fig. 5).

The residual amount of noble gases in the cavity depends on two parameters of the chemical reaction:

- % of the noble gases in the initial mixture
- Initial pressure set point

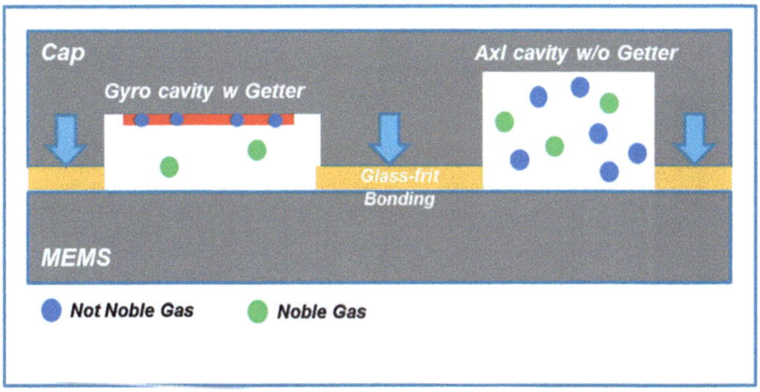

Fig. 5 Working principle of a getter layer

The getter layer helps to maintain a vacuum in the cavity throughout the life of the device.

A key challenge during manufacturing is to build up, control, and guarantee, on the same die, two cavities with substantially different operating pressures: while the Coriolis resonating principle of the gyroscope does require a near-to-vacuum level for performance optimization, the accelerometer on the other side achieves the best performance when the cavity pressure is close to ambient.

After extensive R&D efforts and significant manufacturing experience gained through producing several tens of millions of units for consumer applications, ST is exporting this innovative concept into more demanding automotive applications.

3 The New 6-Axis Solution

Modern navigation systems for the automotive market require accurate and reliable sensors for precise implementation of the dead-reckoning algorithms needed for accurate map positioning in all different driving situations and circumstances.

As with the majority of inertial sensors produced by ST, the ASM330LXH is composed of two dies embedded into an over-molded plastic package.

The single-die processing chain, designed using an advanced silicon technology with 130 nm lithography (HCMOS9A), includes both gyroscope and accelerometer processing chains (Figs. 6 and 7).

Main features of the gyro section are the following:

- The device operates with a closed loop in the driving and an open loop in the sensing chains; both chains are discrete-time to implement precise gains and time constants
- The clock of the whole system is obtained from the well-controlled resonant frequency of the MEMS driving section

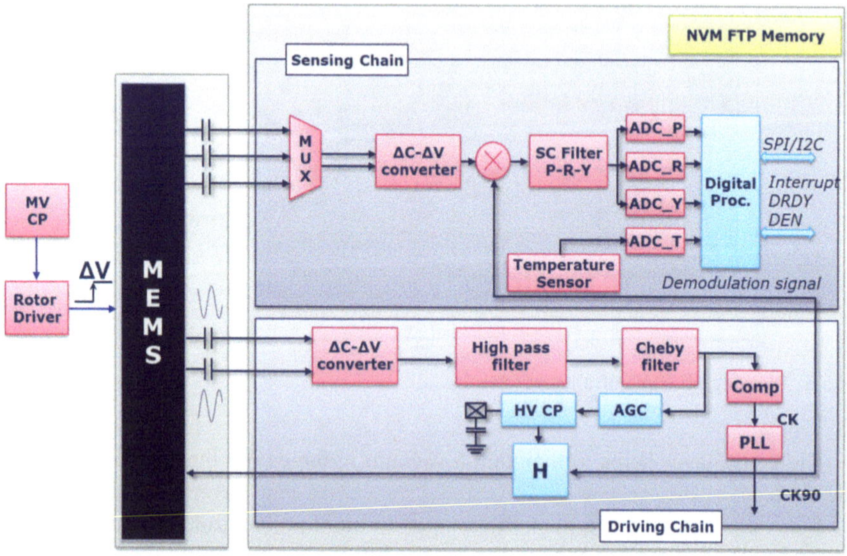

Fig. 6 Block diagram of the gyroscope signal chain

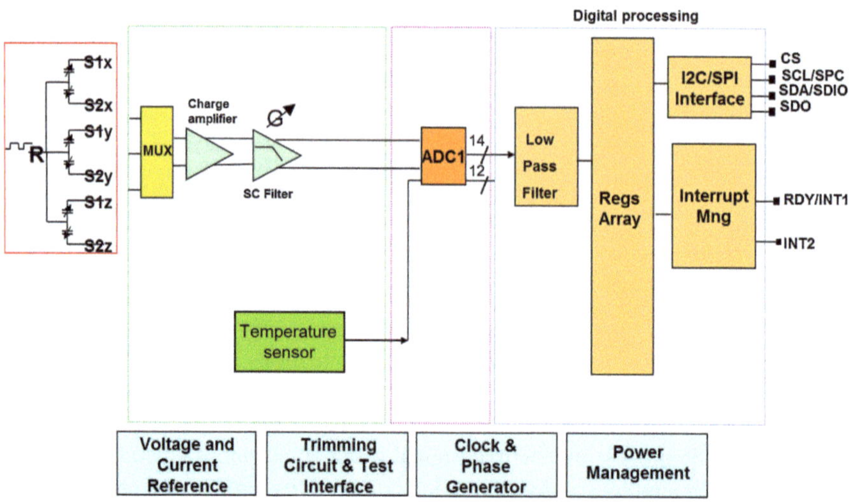

Fig. 7 Block diagram of the accelerometer signal chain

- The input C2V blocks convert the capacitive variation into an electrical signal; the following stages regenerate a continuous time sinusoid from the discrete-time input
- The total phase delay of the closed loop driving satisfies the Barkhausen criteria

- A comparator generates the base clock for the system, which is fed to a Phase Locked Loop
- An Automatic Gain Control stage is used to control the MEMS oscillation amplitude
- The three axes are read out and processed in a time-multiplexed fashion
- The amplitude-modulated input signal is coherently demodulated and then filtered by a switched capacitor filter
- Each axis is processed by a dedicated 14 bit SAR ADC.

Main features of the accelerometer section are the following:

- Time multiplexing approach: one measure chain for all 3 axes
- Charge amplifiers with a Switched Capacitor circuit with CDS
- Fully differential sensing chain from sensor to ADC (all common modes are rejected)
- The chain features a 14-bit fully differential SAR A/D converter, with 25 μs conversion time and low current consumption.

The second die includes both mechanical elements for the 3-axis gyroscope and the 3-axis accelerometer (Fig. 8).

The working principle of the gyroscope is based on the patented "Beating Heart" concept introduced by ST back in 2010 [2, 6], while the low-g accelerometer is based on the traditional comb-finger approach for in-plane sensing and the suspended cantilever approach for out-of-plane sensing.

The device embeds a list of innovative features:

- 3 main operating modes:
  - Only accelerometer active and gyro in power down
  - Both accelerometer and gyro active at the same ODR (synchronized)
  - Both device in power down
  - Switching from one mode to another is done by writing in the device control registers

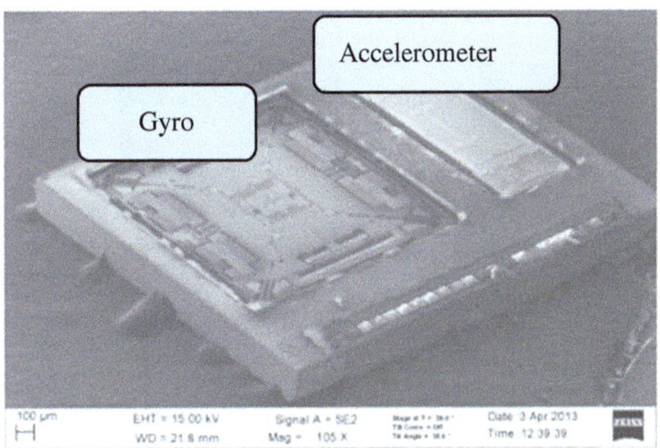

Fig. 8 Dual-in-one sensor

- High level of flexibility granted to the user:
  - Setting of the measurements ranges
  - Output Data Rate selectable
  - Selectable accelerometer anti-aliasing filter
- Enhanced noise performance
- Embedded temperature sensor, with 50 Hz refresh rate and 16LSB/°C temp sensitivity
- Temperature compensation embedded.

The algorithm for temperature compensation, able to reduce offset drifts for both sensors, is of great help in improving the device's performance. The characterization data taken on a sample of 30 units soldered on board show very narrow distributions of key parameters like zero-rate (ZRL, Fig. 9) and zero-g (Fig. 10) levels across the extended temperature range (−40 to +85 °C).

Key advantages offered by ASM330LXH versus other solutions in the market can be summarized hereafter:

- Smallest package for a 6-axis combo for in-dash navigation systems
- High level of silicon integration guarantees a higher reliability level in the field

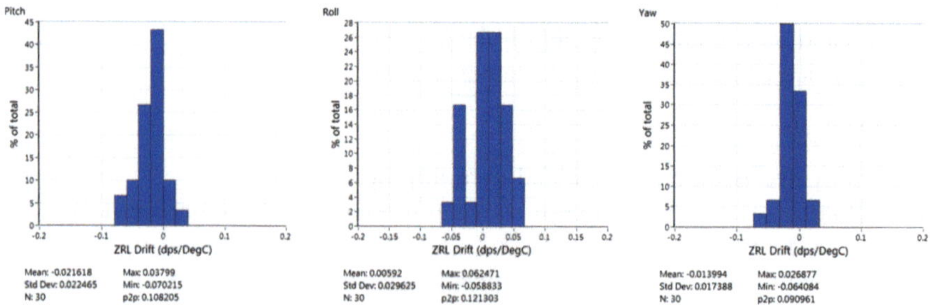

Fig. 9 Gyroscope ZRL offset drift coefficient

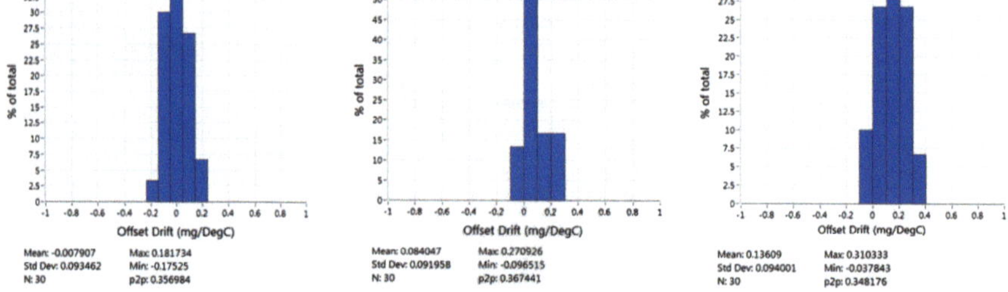

Fig. 10 Zero-g offset drift coefficient

- Dual-in-one approach and single ASIC interface allows perfect sensor output synchronization to achieve the most accurate DR algorithms
- Extreme user flexibility
- High stability in temperature and in time.

4 Trends

The trends in this domain are very exciting.

At ST, one of the first steps is to integrate the ASM330LXH in the TeseoIII positioning platform, as well as enrich the system with additional sensors that could boost overall accuracy, like pressure sensors and electronic compasses.

In general, the traditional navigation market may radically change in the near future with the entry of new companies, who will use their expertise to enable the seamless fusion between portable devices and the currently protected car environment. Internet availability, V2V (Vehicle-to-Vehicle) and V2I (Vehicle-to-Infrastructure) communications will become standard in the coming years.

Moreover, positioning systems will be integrated into a master-system providing key information to other sub-systems dedicated to safety-related tasks like vehicle dynamics and ADAS. Smart tires equipped with sensors will be also part of this Copernican revolution.

5 Conclusions

To guarantee people's safety, autonomous vehicles, which will reach significant diffusion by the mid of the next decade, need sophisticated ADAS systems, along with reliable positioning information with an accuracy below 10 cm, about the size of an orange.

Additional sensors like pressure and electronic compass may be used to achieve this challenging goal.

Thanks to a wide experience in the automotive domain, advanced electro-mechanical technologies, a best in class manufacturing machine, and solid design and engineering experience, ST is correctly positioned to fulfill the forthcoming demand of precise sensors for future systems, which will merge navigation, vehicle dynamics, and ADAS.

References

1. Ferraresi M, Giuffrida G, Palella N (2013) Innovative MEMS sensors in advanced positioning systems. In: Fischer-Wolfarth J, Meyer G (eds) Advanced microsystems for automotive applications 2013. Springer, Cham
2. Ferraresi M, Marinoni A (2012) Tri-axial MEMS gyroscope for automotive applications, SSI conference 2012. ISBN 9783-8007-3423-8

3. Ferraresi M, Pozzi S (2009) MEMS sensors for non-safety applications, AMAA
4. Mattos PG (2011) Accuracy and availability trials of the consumer GPS/GLONASS receiver in highly obstructed environments, IoN GNSS-2011 conference, Portland, Oregon
5. Mattos PG, Pisoni F (2012) Multi-constellation—to receive everything, IoN GNSS-2012 conference, Portland, Oregon
6. Vigna B (2011) STMicroelectronics Masters the Art and the Science of MEMS, Semicon Taiwan
7. Vigna B (2011) Tri-axial MEMS gyroscopes and six degree-of-freedom motion sensors, IEEE, IEDM11-662

Automotive LIDAR-Based Strategies for Obstacle Detection Application in Rural and Secondary Roads

Andrea Carlino, Luciano Altomare, Marco Darin, Filippo Visintainer and Alessandro Marchetto

Abstract The usage of a LIDAR-based obstacle detection system in agricultural environment, as developed in SOLCO project, has led to a novel detection strategy. The *rural* scenario can be rather complex in terms of background, type and size of the obstacles encountered. In the experimentation, a strategy was applied to several trials in rural scenarios, with the aim of finding and fine tuning an object filtering approach that can be later applied in runtime. This strategy could be transferred to passenger car applications to improve LIDAR system performances.

Keywords Object detection · LIDAR · Rural roads · ADAS systems

1 Introduction

ADAS systems are becoming more and more widespread and they support the drivers in a wide numbers of driving scenarios; ADAS functions are usually related to an environment for which some regularity assumptions could be made. Some functions, like Lane Departure Warning (LDW) or Lane centering, in most cases rely on the presence of lines on the road surface (horizontal signage) [1]; other

A. Carlino (✉) · L. Altomare · M. Darin · F. Visintainer · A. Marchetto
Unit of Info-Telematic Systems, Centro Ricerche FIAT – Trento Branch,
Via Della Stazione, 27, 38123 Mattarello (TN), Italy
e-mail: andrea.carlino@crf.it

L. Altomare
e-mail: luciano.altomare@crf.it

M. Darin
e-mail: marco.darin@crf.it

F. Visintainer
e-mail: filippo.visintainer@crf.it

A. Marchetto
e-mail: alessandro.marchetto@crf.it

T. Schulze et al. (eds.), *Advanced Microsystems for Automotive Applications 2015*,
Lecture Notes in Mobility, DOI 10.1007/978-3-319-20855-8_8

functions, like Adaptive Cruise Control, Blind Spot Detection and Cross Traffic Alert, are designed for urban, extra-urban roads or highways, where a crash with obstacles (cars, trucks, motorcycles, bicycles or pedestrians) ahead or crossing should be avoided.

The conditions in rural, agricultural or secondary roads (from now on called simply *rural*) are very different, because of the background, highly variable due to the frequency of curves and slope changes, because of the obstacle type and related features and also due to the different road size and geometry. All these differences influence the precision of the object detection in *rural* scenario. The main drive to this research topic came from the SOLCO project (co-funded by the Public Authority Provincia Autonoma di Trento), where a system both for automatic machinery movement and safety purposes in agricultural environment was designed. One of the main problems experienced was the unstable detection of the objects in the tree rows, due to the high presence of tree trunks and branches without leafs. The LIDAR tended to group "real" objects in fake objects changing in size, position and orientation, failing then the detection of people and real obstacles. Examining the similarities between the specific agricultural and the general *rural* scenario, we designed a processing strategy to manage these detection problems and make the object detection more robust. This strategy could be potentially used also for passenger cars driving in *rural* roads.

In Sect. 2 the main differences between the urban/extra-urban and rural scenarios are presented, both from road and from object features standpoint. In Sect. 3 the system architecture, based on the SOLCO project and simplified for these tests, is described, together with the test conditions. In Sect. 4 the specific strategies for the obstacle detection reinforcement are depicted, while the system evaluation and the experimental results are presented in Sect. 5. Finally, in Sect. 6, the conclusions are drawn and the next tests and possible improvements are discussed.

2 Problem Definition

This chapter introduces the motivation for this research, starting from a general classification of roads, highlighting the main differences between *rural* roads and other kinds of roads and then focusing on the differences related to the obstacle types. The main problems in *rural* scenarios [2], as emerged from the analysis, are the heterogeneity of objects on the way, the multitude of objects in the background and the scarce presence of visible lane boundaries. This influences the obstacle detection and the recognition of the *actual* free path in front of the car.

2.1 Rural and Secondary Road Scenarios

Rural roads are rather heterogeneous, in terms of road characteristics but also of surrounding landscape, as can be seen in Fig. 1.

Examining the several typologies of rural roads, from dirt roads to simple paths within country crops and fields, a number of distinctive features can be highlighted, also by comparing such road types with urban, extra-urban roads and motorways. These differences regard the types of obstacles, the carriageway, the surrounding landscape, which influences the background of the sensor view, the roadside and the road geometry and size. A summary is presented in Table 1.

2.2 Relevant Aspects for Road Perception and Obstacle Detection

Hereafter some highlights of rural roads are given, from the road perception point of view. Road geometry is irregular and often with poor or even no horizontal signage. Lane configuration is generally very simple, with just two lanes or even one single

Fig. 1 Test scenarios

Table 1 Classification of roads and main aspects of each scenario

Feature	Rural scenario	Urban/extra-urban scenario
Obstacles	Very high variability of obstacle size, unpredictable behavior (e.g. wild animal)	Small variability in object classes (cars, pedestrians, bikes, motorbikes, trucks, vans)
Terrain/carriageway	No lane delimitations (horizontal signage), small width even for two-way roads, uneven road surface (holes and irregularities)	Lane delimitations, well defined lanes, regular road surface
Background	Highly variable sensor background due to frequent narrow curves and slope changes	High presence of buildings, more geometric and well-defined shapes
Roadside	No sidewalk, possible close walls, enclosures, few road markings (vertical signage)	Presence of sidewalks or guardrails, many signs (vertical signage)
Geometry	Highly variable width of roads, shorter distances	Standardized and defined width of lanes, larger streets, larger distances

lane. Road markings and other roadside reference points are not a good assumption, because they might be missing in most cases. Field of view can be partially obstructed due to the roadside characteristics or to the presence of curves and slopes.

Concerning obstacle detection, the scenario is very different from the urban case. In front of the vehicle objects could be detected such as plant fronds, which are fake obstacles and can be easily passed through; on the other hand, real obstacles might suddenly appear, such as unexpected pedestrians behind a curve or animals crossing from the roadside. Finally, in narrow two-way roads, two vehicles coming from opposite directions might have to stop and negotiate the way, being thus an obstacle to each other.

In brief, safe driving in rural road is thus not much a matter of lane and other typical “road users” recognition [1], but rather road keeping [3, 4] and general obstacle detection, and reference visible marks are not a good assumption. This fact, combined with the variable light conditions, suggested that LIDAR-based techniques could perform better than a camera system. Furthermore, considering the Horizontal Field Of View (HFOV), the angular resolution and the general flexibility, LIDAR was also preferred over the radar solution. The following section illustrates the prototype that was set-up to test the LIDAR-based obstacle detection in rural roads.

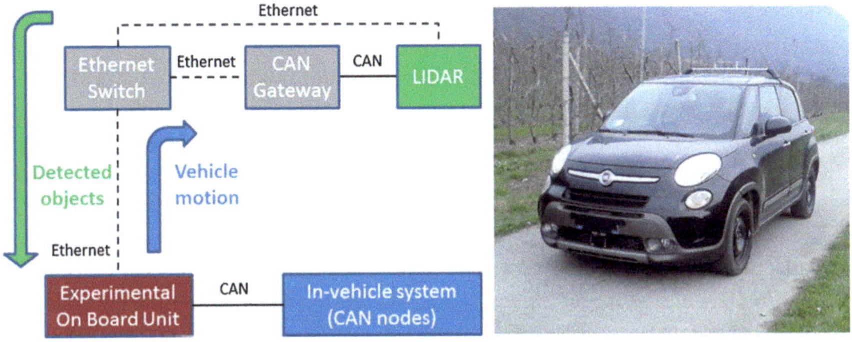

Fig. 2 Scheme of the vehicle architecture and picture of the car prototype

3 Experimental Setup

In order to perform the measurements a vehicle with all the required hardware was setup. The prototype is shown in Fig. 2, along with an architectural scheme. The Experimental On Board Unit converts the vehicle data from vehicle CAN to the LAN network and provides the LIDAR with the required vehicle parameters, used for relative speed and orientation computation. A dedicated CAN Gateway is used to route the vehicle data from the Ethernet cable to the proprietary CAN network of the LIDAR. The LIDAR sensor receives the vehicle data (e.g. speed) and sends a list of detected objects and their related parameters. The On Board Unit processes the LIDAR output for threat assessment.

The device used for obstacle detection is a time-of-flight based 2D Near-Infrared laser scanner for automotive. Its main specifications are listed in Table 2. The most important aspects are the horizontal Field Of View of about 145°, a distance resolution smaller than 100 mm and a range for objects of 80 m.

The detection device is installed on the front bumper of a FIAT 500L, the SOLCO prototype used also for these tests, at a height of 34 cm (ground to vertical center of the sensor) and horizontally centered.

The LIDAR output has been recorded with the proprietary software of the LIDAR manufacturer; each output file includes all the pre-processed data computed

Table 2 Main features of the LIDAR sensor

Feature	Value
Wavelength of sender IR-LD	905 ± 10 nm
Horizontal FOV	~145°
Vertical FOV	3.2° (average)
Horizontal resolution	≥0.25°
Distance resolution	≤100 mm
Range for objects	80 m
Scan rate	25 Hz (1 frame every 40 ms)

by the sensor, such as the vehicle dynamics, the object list along with the relative parameters for each object and the synchronized video from a webcam used as reference base. In the object list each different object is identified from a different object ID; the most interesting parameters for each object are: distance in (x, y) coordinates, in cm, from the center of the LIDAR (frontal center of the car), estimated width and length, orientation, relative and absolute speed and the object age, i.e. the period in milliseconds for which the object has been seen.

The proposed strategy has been tested on the data recorded on six different types of roads, shown above. For each road type three different measures have been recorded, two with the target object to be detected in the center of the carriageway and one with a clear scenario, i.e. without any target object, for comparison. The target object, a pedestrian, was standing still in the center of the available path or carriageway, and the car started to move and record data on the straight way from a distance of 40 m and moving towards the pedestrian at a constant speed of 20 km/h.

4 Obstacle Detection Strategies

The most significant output of the device is a set of objects, each one with its identifier and related attributes, such as relative position, distance in x and y axis and relative speed, as well as its size and, if recognized, classification as, for example, pedestrian or vehicle. The object is tracked so that its trajectory evolution can be monitored and an assessment can be made whether it is going to be an obstacle in the vehicle path or not.

In simple scenarios with few road users or elements of classified type—such as pedestrians and vehicles—and no other objects or physical boundaries in the background, the data processing in runtime is quite simple. Indeed, each single element of the detected object list typically corresponds, both in position and size to the main obstacles as experienced by the driver, and it remains tracked throughout the sensor field of view. In this simplified scenario a threat assessment could be based on real-time trajectory evolution of each single object. However, a scalability problem rises with an increasing number of objects and background complexity.

Sometimes background may get more complex, such as in urban environment, but its regularity (houses, plain walls, fences) allows to clearly identify the scene boundaries which are likely not to be an obstacle and other objects in the middle which may become dangerous.

In limit cases, as may frequently happen when driving through rural roads, the situation is far from simple. The reasons, sometimes combined, are the presence of an unstructured landscape, such as woods and bushes, its high degree of irregularity, as well as the narrow size of the path, which in some cases can be hardly 1 m larger than the vehicle width. In such cases, the LIDAR output is a huge set of objects in the detected field of view. As an order of magnitude, a measurement over 40 m path may sometimes yield up to 1000 objects.

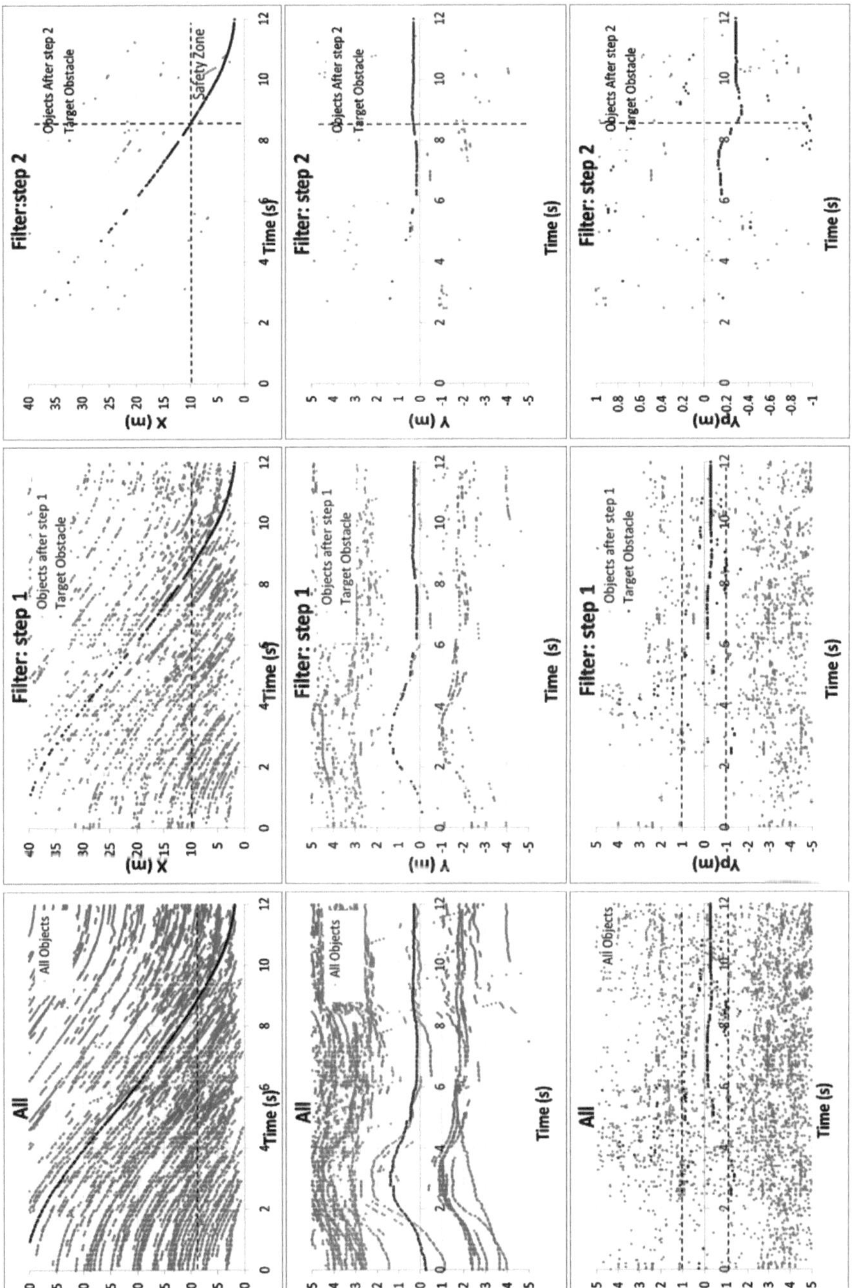

Fig. 3 Filtering steps applied to Rural 3 scenario

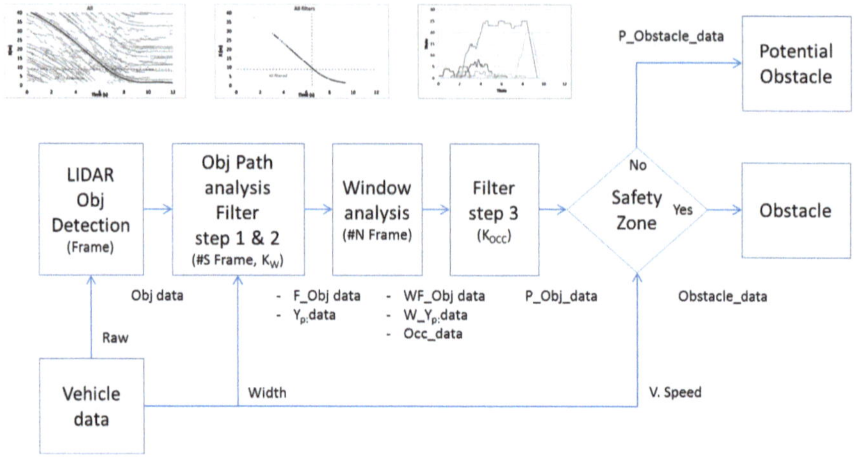

Fig. 4 Representation of the filtering scheme

There are cases where the same object which is lost and tracked again, thus being associated to different identification numbers. Furthermore, it has been experienced how close and irregular road edges may be mapped as large obstacles obstructing the path. In these scenarios, it becomes clear that applying a plane post processing threat for every detected object is not a scalable solution and fast filtering has to be done in advance, in order to focus on the real and closer threats. The strategy here proposed is based on an a posteriori analysis of collected data on a collinear obstacle detection in rural scenarios. Figure 3 shows, for the sample Rural 3 scenario, the time behavior of object longitudinal distance from the car (X), lateral distance (Y) and lateral trajectory projection on the car frontal plane (Yp). Figure 4 represents schematically the step by step data analysis that was performed in order to identify potential obstacles and optimize noise filtering.

Firstly, a two-step object path analysis and filtering was performed based on subsequent values of trajectory. The first step analyses subsequent frames and filters out object detections whose trajectories are not stable. The second step determines whether filtered object trajectories are conflicting with the ego-vehicle cross-section, considered as 2 m (width) in the experiments. The result is a filtered set in which target obstacles are identifiable by a pattern persistence (Fig. 3, second step filter), but still sporadic detections exist which are not the target obstacle. However, experiments showed that even in complex scenarios, the occurrences of such objects are generally much less than the target obstacle. The solution proposed is based on an additional noise cleaning through a time-windowed filter over the number of occurrences with the same object identifier.

Finally a threat assessment is done, based on the time to brake. In rural scenarios, where tests were performed at 20 km/h (5.6 m/s), a conservative longitudinal safety area was considered to be 10 m, given that typical braking distances at this speed are 2–3 m with dry pavement and from 5 to 10 m with wet to extremely slippery ground.

The following section shows the first experimental results obtained by applying this strategy in the different scenarios.

5 Preliminary Results

Figure 5 reports the target obstacles after the two-step filtering, for all 6 scenarios with 40 m initial target distance. As expected, the distance of potential obstacle recognition after filtering out the noise, decreases with increasing scenario complexity. For example comparing the Extra-Urban with the Fields 2 scenario (see also Fig. 1), this distance decreases from 30 to 20 m, but in the field scenario the object is also seen intermittently between 10 and 20 m.

The graphs in Fig. 6 show the last step of window-based filtering on the number of occurrences, for three representative scenarios (Extra-urban, Rural 3 and Fields 2). The threshold of 7 occurrences in the time window was applied, and then the

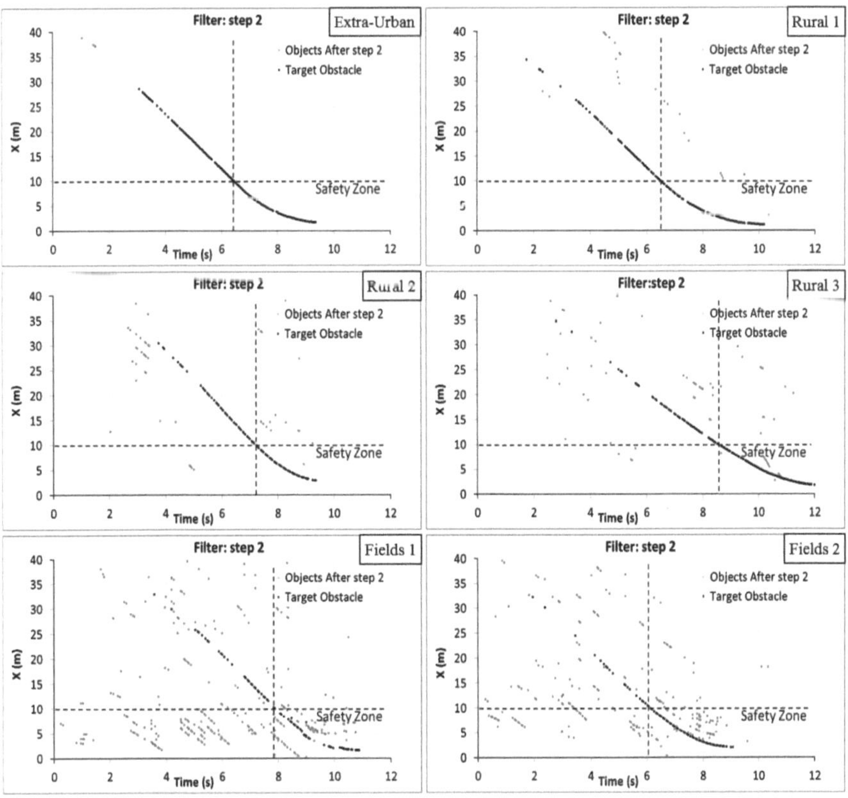

Fig. 5 Two-step filtering results for all scenarios

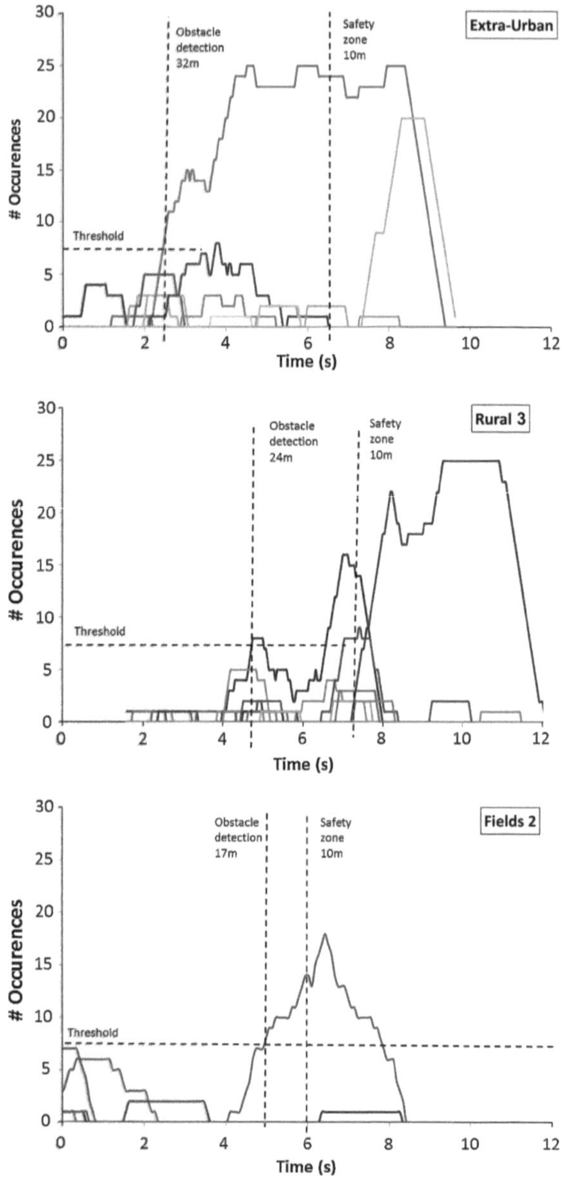

Fig. 6 Time-windowed filter over the number of occurrences

remaining object identifiers were confronted with the target obstacle identifiers. The results show that in all three scenarios the target obstacle was distinguished from other objects, at decreasing distance when going to more complex landscape, but before the safety zone.

6 Conclusions and Next Steps

A basic difference in the road condition exists between the rural and secondary roads and the urban, extra-urban roads and highways; the work presented in this paper investigated the main differences that can afflict the obstacle detection based on a LIDAR sensor and proposed an approach based on subsequent filters applied on the LIDAR output. These filters, based on object trajectory stability, on intersection with the vehicle path, on detection persistence and on safety area definition, proved to be effective in cleaning the noise in the LIDAR detected object list and in extracting the real threats in rural scenarios.

In order to consolidate the proposed strategy more testing configurations should be considered. As a starting point there are five key variables to take into account. The first one is the obstacle size, which in rural scenarios can be smaller or larger than a pedestrian. The second one is the obstacle position, to be moved from the center to the side of the road; in addition, the possible dynamic states of the obstacle can play an important role, i.e. whether the target is moving and if on a regular or irregular path (sudden change of direction). The last two variables, which should be increased, are the speed of the ego vehicle and the distance of the target object.

The proposed strategy seems promising and is currently being implemented and tested in a real-time software application. The next challenge is to find a compromise between reliability and processing delay, in view of future deployment on passenger cars.

Acknowledgments The field testing presented in this research work was supported by Fondazione Edmund Mach (FEM), which allowed us to perform the tests in the tree lines of its farm. We also want to thank all the SOLCO team and in particular our colleague Gianluca Demattè who provided his expertise in the data analysis that greatly helped achieving the results.

References

1. Hillel AB, Lerner R, Levi D, Raz G (2010) Recent progress in road and lane detection: a survey, machine vision and applications. Springer, Berlin
2. Choi J, Lee J, Kim D, Soprani G, Cerri P, Broggi A, Yi K (2012) Environment-detection-and-mapping algorithm for autonomous driving in rural or off-road environment. IEEE Trans Intell Transp Syst 13(2):974–982
3. Zhang W (2010) LIDAR-based road and road-edge detection. In: IEEE intelligent vehicles symposium, pp 845–848
4. Takagi K, Morikawa K, Ogawa T, Saburi M (2006) Road environment recognition using on-vehicle LIDAR. In: Intelligent vehicles symposium, pp 120–125
5. Zindler K, Geiß N, Doll K, Heinlein S (2014) Real-time ego-motion estimation using lidar and a vehicle model based extended Kalman filter. In: IEEE 17th international conference on intelligent transportation systems (ITSC), pp 431–438
6. Shim I, Choi D-G, Shin S, Kweon IS (2012) Multi Lidar System for fast obstacle detection. In: 9th international conference on ubiquitous robots and ambient intelligence (URAI), pp 557–558

7. Sivaraman S, Trivedi MM (2013) Looking at vehicles on the road: a survey of vision-based vehicle detection. Tracking Behav Anal IEEE Trans Intell Trans Syst 14(4):1773–1795
8. Mammeri A, Zhou D, Boukerche A, Almulla M (2014) An efficient animal detection system for smart cars using cascaded classifiers. in: IEEE ICC 2014—CSSMAS, pp 1854–1859
9. Shang E, An X, Li J, He H (2014) A novel setup method of 3D LIDAR for negative obstacle detection in field environment. In: IEEE 17th international conference on intelligent transportation systems (ITSC), pp 1436–1441

Part III
Networked Vehicles, ITS and Road Safety

Performance Evaluation of a Novel Vehicle Collision Avoidance Lane Change Algorithm

Sajjad Samiee, Shahram Azadi, Reza Kazemi, Arno Eichberger, Branko Rogic and Michael Semmer

Abstract This study, proposes a methodology to evaluate the performance of a novel emergency lane change algorithm. The algorithm, defines a number of constraints, based on the vehicle's dynamics and environmental conditions, which must be satisfied for a safe and comfortable lane change maneuver. Inclusion of the lateral position of other vehicles on the road, the tire-road friction, and real-time ability are the main advantages of the proposed algorithm. For performance evaluation of the developed algorithm, a set of driving scenarios were designed to consider different possible traffic situations that may appear in an emergency lane change maneuver. These scenarios were implemented later in IPG CarMaker, which is a vehicle's dynamics platform. Based on the designed scenarios, the efficiency of the algorithm in collision free lane change maneuver was examined.

Keywords Autonomous driving · Lane change maneuver · Decision making · Collision avoidance

S. Samiee (✉) · S. Azadi · R. Kazemi
K.N. Toosi University of Technology, No. 15, Pardis Street, Vanak Square, Tehran, Iran
e-mail: s.samiee@tugraz.at

S. Azadi
e-mail: azadi@kntu.ac.ir

R. Kazemi
e-mail: kazemi@kntu.ac.ir

S. Samiee · A. Eichberger · B. Rogic · M. Semmer
Graz University of Technology, Inffeldgasse 11, Graz 8010, Austria
e-mail: arno.eichberger@tugraz.at

B. Rogic
e-mail: branko.rogic@tugraz.at

M. Semmer
e-mail: michael.semmer@tugraz.at

T. Schulze et al. (eds.), *Advanced Microsystems for Automotive Applications 2015*,
Lecture Notes in Mobility, DOI 10.1007/978-3-319-20855-8_9

1 Introduction

Cars have a great impact on our life. They are symbols of freedom and are often used as a means of self-expression. But they can change our life to the worst. In the year 2010, 28.759 people died in the Europe [1] and 32.999 in the USA [2] because of car accidents. The lane change maneuver, is one of the serious cause of car accidents especially as a consequence of drivers' errors on accurate estimation of the distance between vehicles [3]. According to the statistics, twenty percent of highway car accidents are a result of an inappropriate lane change [4]. Also, lane change is a frequent driving maneuver and considered to be as a next step in automation of driving. In addition, it is an important part of microscopic traffic simulation and has a considerable effect on analysis results of these models [5]. Given all the aforementioned reasons, autonomous lane change is an important topic of study in automotive engineering nowadays.

Various methods have been demonstrated for lane change decision making and path planning so far. In [6], a model was developed for vehicles lane change based on the cellular automaton (CA), which mainly focused on some of the vehicle's constraints such as maximum acceleration and deceleration. The rules used in [6] were later used in another study for traffic simulation in double- and triple-lane broad highways [7]. It was demonstrated that the developed model allows realistic simulations. A soft computing method was used in this study to model driver behavior during the lane change. In order to have more comprehensive model which covers complicated scenarios, the proposed system had more than only a single input and single output (SISO) [8].

In another study, an algorithm was proposed which was able to identify the boundaries of the path, store the obtained information and design the desirable driving path using a vectorial approach [9]. In [10], the driving task was interpreted as a model predictive control which was able to control and stabilize double-lane change maneuver using fuzzy logic in accordance to the ISO standard. The aforementioned approach was also employed in another study to control vehicle velocity in addition to the lane change maneuvering [11]. The experiments conducted on a one-way two-lane road demonstrated suitable longitudinal and lateral control action of the vehicle consistent with the traffic condition of the road.

In addition to the studies about decision making, a lot of studies are focused on path planning technics. Some models where developed based on the vehicle's dynamics and the driver strategy during the lane change maneuver. The experimental evaluations during real driving conditions showed that the proposed models are superior to those using polynomial for path planning and produce more accurate paths [12, 13]. Intelligent control techniques, such as fuzzy control [14], neural networks [15] and swarm intelligence [16], were also employed for path planning. For instance, neural networks were employed in [17] to predict movements of the other vehicles in short- and long-time. Long-term predictions were used to warn the driver to do lane change or avoid it due to the possibility of collision. In addition, short-term predictions helped the driver to deal with unexpected changes in a traffic flow.

2 Algorithm

To develop the lane change decision making algorithm, first, the equations for the lateral movement of the vehicle in terms of maneuver time are produced. Then, the critical maneuvering time is calculated on the basis of the constraints. Finally, the feasibility of carrying out the maneuver is decided upon by comparing the critical times. It is assumed that, in the worst-case scenario there are three other surrounding vehicles during the maneuver, as shown in Fig. 1. Vehicle E represents the ego (lane changer) vehicle, vehicle A represents the leading vehicle at the same lane, and vehicles B and D are leading and rear vehicles at the target lane, respectively. Moreover, the dashed-line vehicle in Fig. 1 indicates the vehicle E during the maneuver. If the four conditions below are satisfied, the lane change maneuver will be possible;

1. During the maneuver, the lateral distance between the right front corner of vehicle E and right rear corner of vehicle B must be at least C_1 (Fig. 1a).

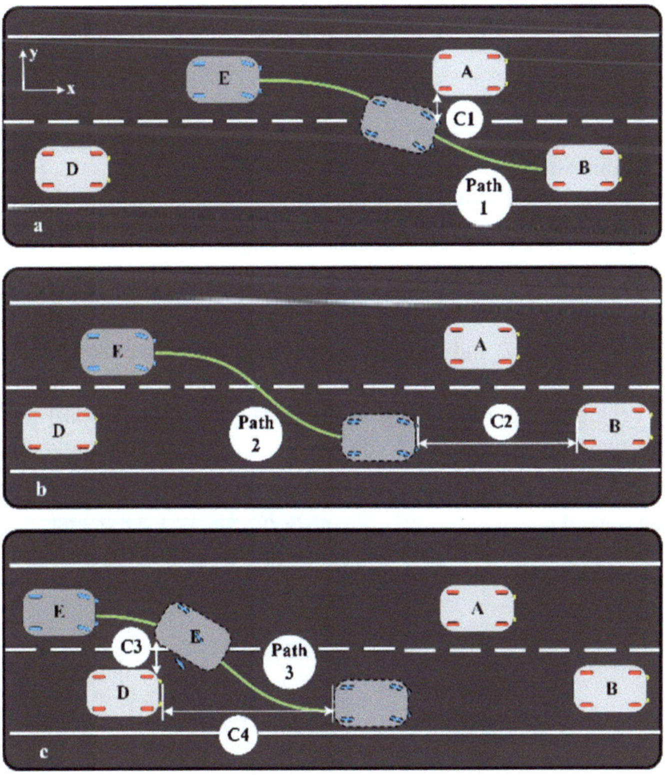

Fig. 1 Definition of constraints in lane change maneuver

2. After the maneuver and movement of vehicle E to the target lane, it distance from vehicle B must be C_2 (Fig. 1b).
3. During the maneuver, the lateral distance from right rear corner of vehicle E to the left front corner of vehicle D must be at least C_3. Moreover, after the maneuver is done the longitudinal distance between these vehicles must be at least C_4 (Fig. 1c).
4. The generated lateral acceleration of E during the maneuver must be achievable, considering the prevailing friction potential between the road and tire.

The proposed decision-making algorithm investigates the possibility of designing a trajectory, taking all abovementioned constraints into account. It focuses on time as the main decision-making parameter. First, the lane change duration for the most critical trajectory in terms of each constraint, is derived. Then, the lane change possibility is decided upon by comparing the computed lane change durations. In the following, the methodology of calculating critical trajectories based on each of the aforementioned constraints, will be described.

2.1 Case 1: A Vehicle in Front on the Same Lane

Considering Fig. 1a, during the lane change the left front corner of vehicle E (point *P*) will touch the right rear corner of vehicle *A* (point *M*) if C_1 is zero. Magnified illustration of this situation is shown in Fig. 2.

Considering the safe distance of C_1 between the vehicles when their longitudinal coordinates coincide, one will obtain Eq. (1).

$$y_A(t) - y_E(t) = C_1 + |O_A M| \sin(\theta_M - \theta_A(t)) + |O_E P| \sin(\theta_P - \theta_E(t)) \quad (1)$$

In Eq. (1), $y_A(t)$ and $y_E(t)$ indicate the lateral position of the center of gravity of vehicles *A* and *E*, respectively. $|O_E P|$ is the length of the imaginary line connecting vehicle *E*'s center of gravity to point *P*. Similarly, parameter $|O_A M|$ indicates the length of the imaginary line between vehicle *A*'s center of gravity and point *M*. θ_M is

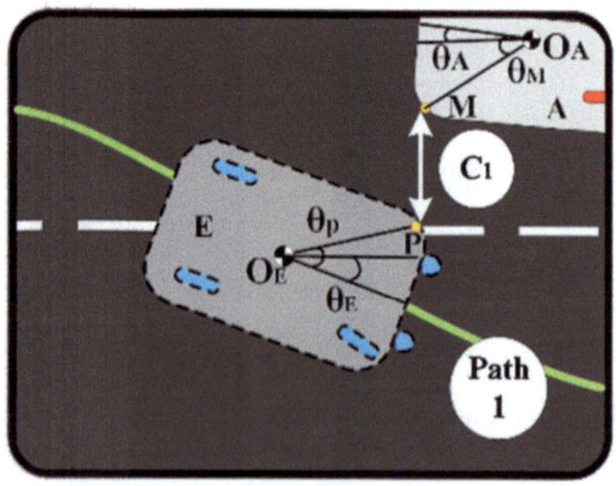

Fig. 2 Lateral constraint between ego vehicle and vehicle in front on the same lane

the angle between O_AM and longitudinal axis of the vehicle A. Parameter $\theta_A(t)$ is the angle between vehicle A's longitudinal axis and the horizon and θ_P is the angle between O_EP and longitudinal axis of the vehicle E. while $\theta_E(t)$ represents the angle between the longitudinal axis of vehicle E and the horizon at any moment. Using the numerical technique presented in [18], one can solve (1) and obtain the maneuver duration such that constraint C_1 is satisfied. This time is labelled as t_1.

2.2 Case 2: A Vehicle in Front and on the Target Lane

Various studies have addressed the issue of the minimum safe longitudinal distance between two vehicles and several formulations have been developed for this distance, e.g. [19, 20]. In this study, the method proposed by Juala et al. [21] is employed. In this conservative method, it is assumed that the velocity of the front vehicle suddenly becomes zero in case of collision with an obstacle. In this circumstance, the safety distance is obtained as,

$$C_2 = s_0 + v_{xE}t_d + \left(v_{xE}^2/2a_{Eb}\right) \tag{2}$$

In (2), s_0 is the safe stopping distance, while a_{Eb} is the maximum deceleration of vehicle E. In addition, t_d is the reaction time of the driver which depends on various factors such as physical and mental condition of the driver as well as road conditions and usually varies between 0.67 and 1.11 [22]. By substituting all required parameters in (2), C_2 and hence the maneuver time, labeled as t_2, can be obtained. Hence, at the specified time instant, the longitudinal and lateral position of two vehicle are governed by (3) and (4);

$$x_B(t) - x_E(t) = s_0 + v_{xE}t_d + \left(v_{xE}^2/2a_{Eb}\right) + l_{Ef} + l_{Br} \tag{3}$$

$$y_B(t) = y_E(t) \tag{4}$$

where, $x_B(t)$ and $y_B(t)$ indicate the longitudinal and lateral positions of the center of gravity of vehicle B respectively, and l_{Br} indicates the longitudinal distance from vehicle B's center of gravity to the vehicle's rear. Obviously, this constraint designates all trajectories in which the longitudinal distance between centers of gravity of vehicles E and B at the end of the maneuver is greater than the value obtained in (3), as a candidate for a safe trajectory.

2.3 Case 3: A Vehicle Behind and on the Target Lane

This case is a combination of the first two cases. A larger illustration of the vehicles condition in this case, is shown in Fig. 3.

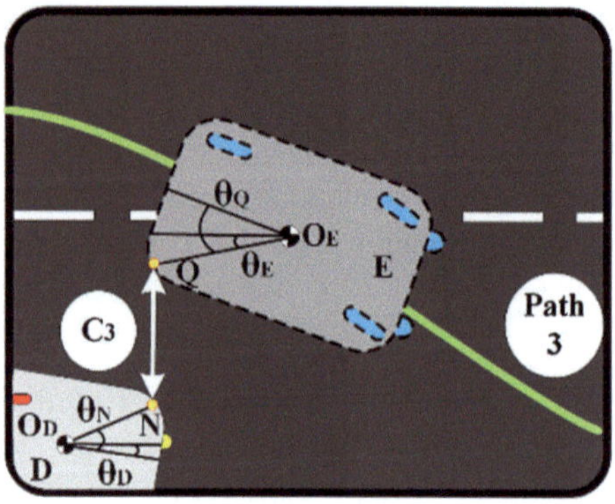

Fig. 3 Lateral constraint between ego vehicle and vehicle behind on the target lane

To obtain the lane change duration, firstly the appropriate maneuver time is computed based on the safe lateral distance using (5). Then, the suitable maneuver time is obtained using the safe longitudinal distance at the end of the maneuver using (6). As the behavior of vehicle D is controlled by the automatic system, the possibility of sudden velocity change is almost negligible and hence a two-second law [23] is used instead of the conservative method in case two. Finally, the larger value among the two obtained values are introduced as t_3.

$$y_E(t) - y_D(t) = C_3 + |O_DN|\sin(\theta_N - \theta_D(t)) + |O_EQ|\sin(\theta_E(t)) \tag{5}$$

$$x_E(t) - x_D(t) = 2v_{xD} + l_{Er} + l_{Df} \tag{6}$$

In Eqs. (5) and (6), $x_D(t)$ and $y_D(t)$ represent the longitudinal and lateral position of the vehicle D's center of gravity. Moreover, v_{xD} and l_{Df} indicate the longitudinal velocity of vehicle D and the longitudinal distance from vehicle D's center of gravity to the vehicle's back respectively. $|O_EQ|$ is the length of the imaginary line between vehicle E's center of gravity and right rear corner of the vehicle, i.e. point Q. Similarly, $|O_DN|$ in (5) show the length of the imaginary line from the gravity center of vehicle D and its left front corner (Point N). l_{Er} is the longitudinal distance between vehicle E's center of gravity and its rear and θ_N indicates the angle between this line and longitudinal axis of the vehicle.

2.4 *Case 4: The Most Aggressive Lane Change*

The designed lane change trajectory for the vehicle must be feasible with respect to vehicle dynamics. In other words, in addition to continuity and differentiability of the trajectory, the dynamic constraints of the vehicle must be satisfied. In particular, it must be ensured that the generated lateral acceleration during the maneuver must be attainable, considering road-tire friction, and maintain vehicle stability. The dynamic vehicle simulation tool, IPG CarMaker, is used for the analyses. Figure 4

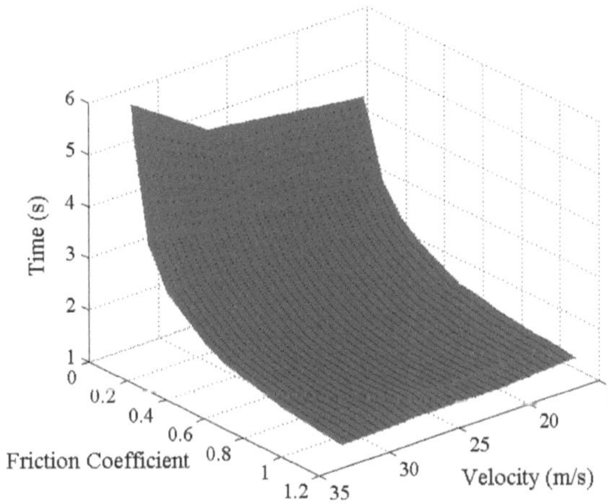

Fig. 4 Diagram of the maneuver time in terms of weight, velocity and road-tire friction

shows the 3-D diagram of the maneuver time in terms of mass, velocity and tire-road friction. The illustrated surface divides the space into two parts. The volume above the surface indicates acceptable maneuver time.

The results of the different simulations are approximated by (7), where the minimum maneuver time t_m is expressed in terms of road-tire friction μ and vehicle velocity v_x.

$$t_m(\mu, v_x) = (\mu(8 + 0.5v_x) + 5)/10\mu. \tag{7}$$

2.5 Decision-Making Strategy

In this paper, the value of the required parameters to obtain t_1, t_2 and t_3 based on the aforementioned equations, are presented in Table 1.

By calculation and comparison of these times, the decision can be made. Table 2 presents the possible lane change cases along with the acceptable time or time interval for the maneuver. Obviously, if the comparison of the computed time does not correspond to any of the cases presented in Table 2, the lane change maneuver is not allowed. If any of the three vehicles *A*, *B* and *D* does not exist on the path, its corresponding time is eliminated from calculations.

3 Scenario Design

Comprehensive scenarios including all different possible traffic situations are designed to evaluate the performance of the developed decision making algorithm. As mentioned earlier, the algorithm is designed to bring the target vehicle to the

Table 1 Value of the parameters

Parameter	Value	Parameter	Value
C_1	1 (m)	w_A	1.65 (m)
l_{Er}	1.9 (m)	l_{Df}	1.7 (m)
w_E	1.56 (m)	S_0	2 (m)
t_d	0.5 (s)	a_{Eb}	0.7 g (m/s^2)
l_{Ef}	1.6 (m)	l_{Br}	1.8 (m)

Table 2 Decision making table

No.	Case	Time	No.	Case	Time
1	$t_1 > t_2 > t_4 > t_3$	$[t_2\ t_4]$	8	$t_2 > t_1 = t_3 > t_4$	t_1
2	$t_1 > t_2 > t_3 > t_4$	$[t_2\ t_3]$	9	$t_1 > t_2 = t_4 > t_3$	t_2
3	$t_2 > t_1 > t_4 > t_3$	$[t_1\ t_4]$	10	$t_1 > t_2 = t_3 > t_4$	t_2
4	$t_2 > t_1 > t_3 > t_4$	$[t_1\ t_3]$	11	$t_2 > t_1 > t_4 = t_3$	$[t_1\ t_3]$
5	$t_1 = t_2 > t_4 > t_3$	$[t_1\ t_4]$	12	$t_1 > t_2 > t_4 = t_3$	$[t_2\ t_3]$
6	$t_1 = t_2 > t_3 > t_4$	$[t_1\ t_3]$	13	$t_1 = t_2 = t_3 > t_4$	$[t_1\ t_3]$
7	$t_2 > t_1 = t_4 > t_3$	t_1	14	$t_1 = t_2 = t_3 = t_4$	t_1

right line of the road only if the driver is not able to control the vehicle. So, the lane change direction will always be to the right.

In the designed scenarios, the target vehicle is moving in a three lane highway, and in the worst case, surrounded by three other vehicles. The scenarios are designed based on the number of vehicles on the road, relative distance, velocity and acceleration of all vehicles, and road friction. For a more clear presentation of different possible traffic situations, the scenarios are defined parametric as it is demonstrated in the Table 3.

3.1 Scenario 1: No Other Vehicle on the Road

The simplest situation happens when there is no other vehicle on the road. The decision making unit defines a proper time for lane change based on the velocity of the ego vehicle and road condition. This situation is presented in Fig. 5.

Table 3 Parametric presentation of different possible traffic conditions

Vehicle	Relative Dist. (m)		Vel. (m/s)		Accel. (m/s^2)		Friction
	Variable	Range	Variable	Range	Variable	Range	Range
Ego	–	[0 60]	v_E	[0 36]	a_E	[−6 4]	[0.1 1.2]
Target 1	s_1		$v_1 = v_A$		$a_1 = a_A$		
Target 2	s_2		$v_2 = v_B$		$a_2 = a_B$		
Target 3	s_3		$v_3 = v_D$		$a_3 = a_c$		

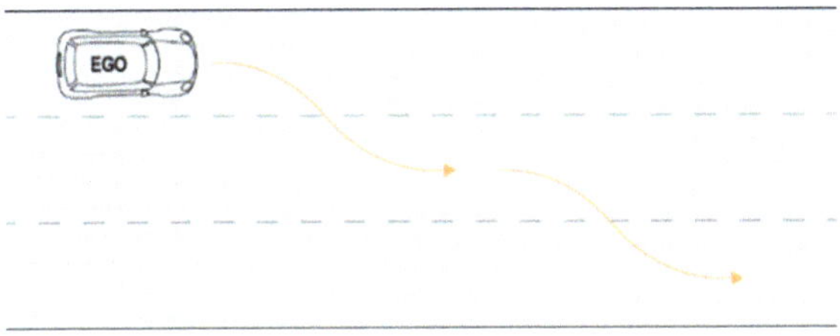

Fig. 5 The simplest possible scenario, no other vehicle on the road

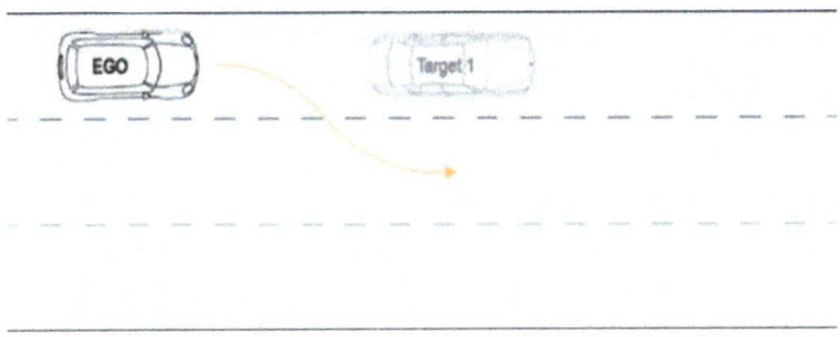

Fig. 6 One vehicle in front and on the same lane

3.2 *Scenario 2: A Vehicle in Front on the Same Lane*

This situation happens when there is only one vehicle in front and on the same lane. Based on vehicle *A* (target 1) dynamic behavior, decision making algorithm must be able to guide the ego vehicle to a safe lane change (Fig. 6).

3.3 *Scenario 3: One Vehicle on the Same Lane and One on Target Lane*

In this case, the decision making system should consider the situation and behavior of two other vehicles in order to prepare a safe lane change maneuver. Here, in addition to the Vehicle *A* which is driving on the same lane as the ego vehicle, The vehicle *B* (target 2) is driving on the target lane. This situation is presented in Fig. 7.

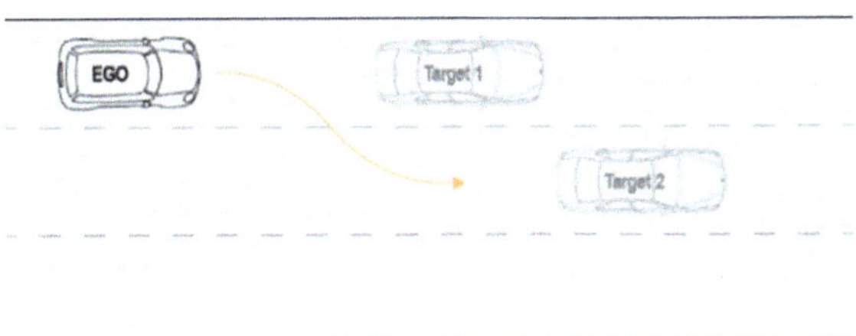

Fig. 7 One vehicle on the same lane and one on the target lane

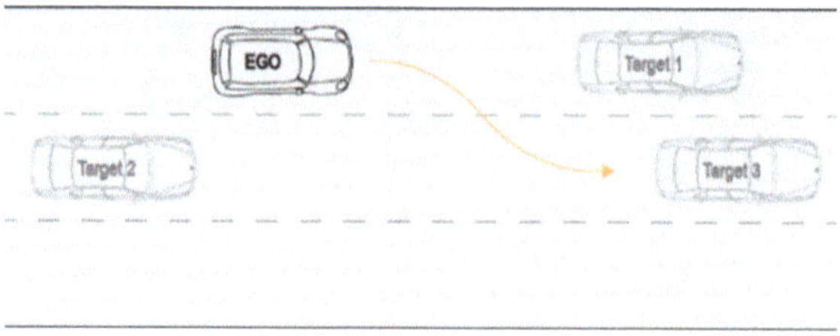

Fig. 8 One vehicle on the same lane and two on the target lane

3.4 Scenario 4: One Vehicle on the Same Lane and Two on Target Lane

This is the most complicated situation for an emergency lean change maneuver where the decision making unit should guide the ego vehicle considering three other vehicles on the road. This situation is presented in Fig. 8.

4 Implementing Scenarios in IPG CarMaker

All the aforementioned scenarios were implemented in IPG Carmaker which is a platform to simulate vehicle's dynamics and control units. To do so, the ability of this software to communicate with MATLAB/SIMULINK was applied for better flexibility. Different constraints in decision making unit, as discussed earlier, were implemented to find out appropriate time t_1 to t_4 in each case. Based on calculated times, the possibility of performing a safe lane change and the corresponding maneuver time were reached based on decision making rules Table 2. Figure 9 shows the block diagram of final system in Simulink.

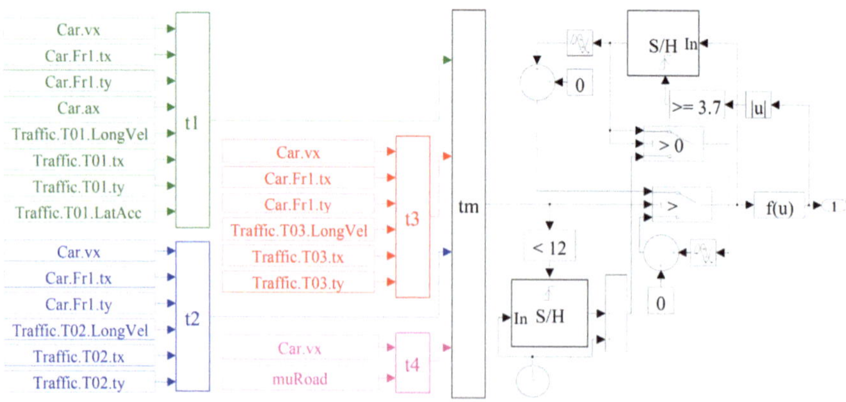

Fig. 9 The Block diagram of decision making system in MATLAB/Simulink

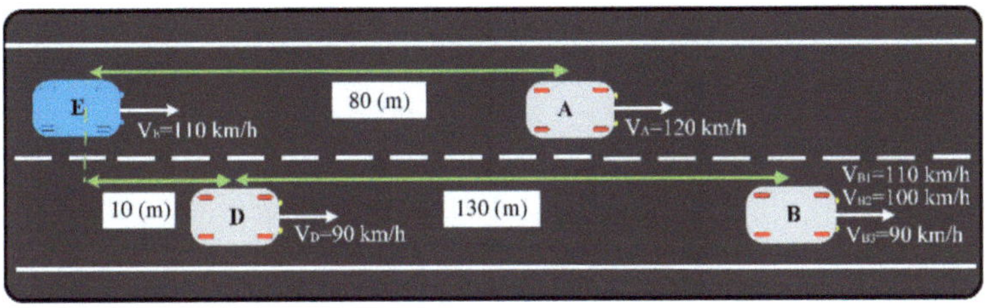

Fig. 10 Sample scenario used to evaluate lane change algorithm performance

In Fig. 9, dynamic data of the ego and target vehicles as well as tire-road friction is fed into the functions where the critical lane change times t_1, t_2, t_3, and t_4 are calculated based on equations presented in Sect. 2. These times are then sent to the decision making function where the rules demonstrated in Table 2 are coded in order to find the final maneuver time (t_m) which is finally presented to the driver model via port number 1. The driver model is responsible to guide the vehicle on the lane change trajectory using Eq. (8).

$$y_E(t) = \left(-6h/t_m^5\right)t^5 + \left(15h/t_m^4\right)t^4 + \left(-10h/t_m^3\right)t^3 \tag{8}$$

In Eq. 8, h is the maximum lateral displacement of the ego vehicle at the end of the maneuver and attains the value of −3.75 which is the standard lane width. The negative sign indicates lane change to the right-side of the road.

5 Results

To evaluate the performance of the decision making algorithm, various tests were performed based on aforementioned scenarios in Sect. 3. In this paper, the results of three different cases are presented and discussed. In case one, the ego vehicle velocity is v_E = 110 (km/h) and vehicle A is driving with the speed of v_A = 120 (km/h) and 100 (m) away in front. Also, vehicles B and C are driving in target lane with the speed of v_{B1} = 110 (km/h) and v_C = 90 (km/h) and are located at the distances of 140 (m) and 10 (m) in front of ego vehicle, respectively. In case two and three, vehicle B is driving at the speed of v_{B2} = 100 and v_{B3} = 90 km/h respectively while all other conditions remain the same as in case one. All explained cases are presented schematically in Fig. 10. Tire-road friction is equal to 0.9 for all cases and the lane change maneuver begins 4.5 (s) after the scenario begins.

Table 4 shows critical maneuver times based on Eqs. 1–7 and final maneuver time based on decision making rules in Table 2. As it can be seen, for the first two cases, the decision making unit allows the lane change maneuver to be performed.

Table 4 Allowable maneuver times for each three cases

Case	v_b (m/s)	μ	t_2 (s)	t_3 (s)	t_4 (s)	t_m (s)
1	30.6	0.9	4.2	1.26	3.2	[3.2 4.2]
2	27.8	0.9	3.6	1.26	3.2	[3.2 3.6]
3	25	0.9	3	1.26	3.2	–

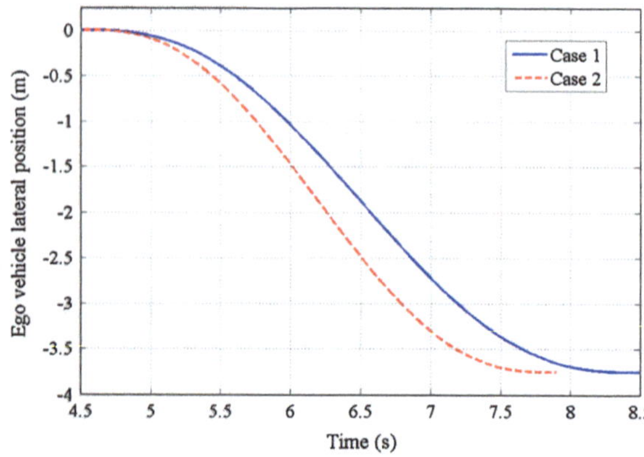

Fig. 11 Lateral position of ego vehicle

The maneuver time for case 1 must be in the range between 3.2 (s) and 4.2 (s) and for case 2 between 3.2 (s) and 3.6 (s). The lane change cannot be performed in case 3.

The diagrams of lateral position of ego vehicle is demonstrated in Fig. 11 for cases one and two. In addition, corresponding lateral velocity and acceleration of the ego vehicle is presented in Figs. 12 and 13. As it can be seen in both cases, the lateral acceleration value is less than 2 (m/s^2) which satisfies passenger comfort condition.

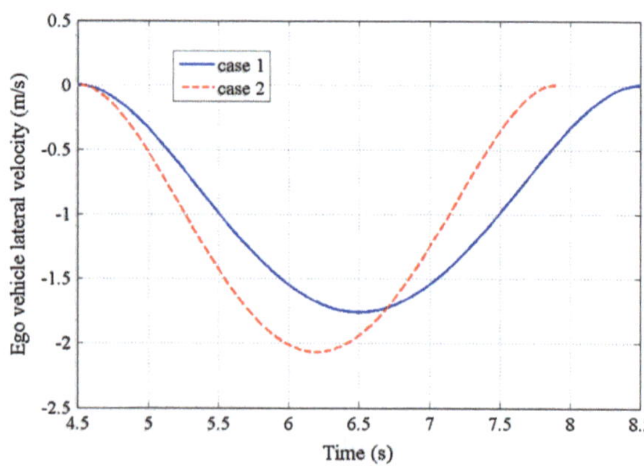

Fig. 12 Lateral velocity of ego vehicle

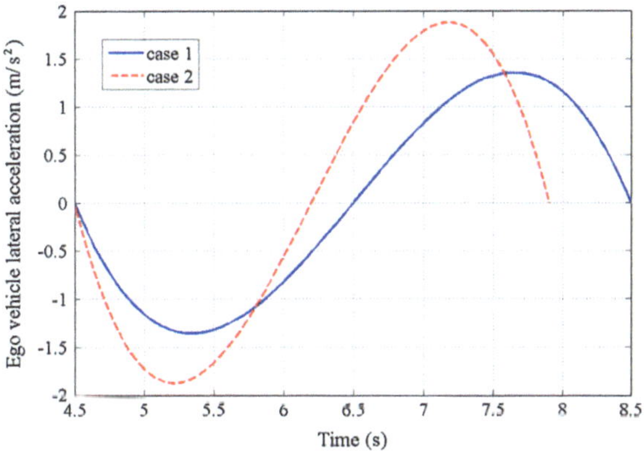

Fig. 13 Lateral acceleration of ego vehicle

6 Conclusion

This study, proposed a methodology to evaluate the performance of a novel emergency lane change algorithm. Inclusion of the lateral position of other vehicles on the road, the tire-road friction, and real-time ability are the main advantages of this novel algorithm. For performance evaluation of the developed algorithm, a set of driving scenarios were designed to consider different possible traffic situations that may appear in an emergency lane change maneuver. These scenarios were implemented later in IPG CarMaker, which is a vehicle's dynamics platform. The result of three different cases were presented and discussed. The results show an acceptable performance of the algorithm.

References

1. Broughton J, Brandstaetter C, Yannis G, Evgenikos P, Papantoniou P, Candappa N, Christoph M, van Duijvenvoorde K, Vis M, Pace JF, Tormo M, Sanmartín J, Haddak M, Pascal L, Amoros E, Thomas P, Kirk A, Brown L (2012) Assembly of annual statistical report and basic fact sheets. Tech. Rep. Deliverable D3.9 of the EC FP7 project DaCoTA, 2012
2. Blincoe LJ, Miller TR, Zaloshnja E, Lawrence BA (2010) The economic and societal impact of motor vehicle crashes. National Highway Traffic Safety, Washington, DC, Tech. Rep. DOT HS 812 013, 2010
3. Djenouri D, Soualhi W, Nekka E (2008) VANET's mobility models and overtaking: an overview. In: 3rd international conference on information and communication technologies: from theory to applications (ICTTA), pp 1–6
4. Wang J, Chai R, Wu G (2014) Changing lane probability estimating model based on neural network. In: 26th Chinese control and decision conference (CCDC), pp 3915–3920
5. Mathew TV (2014) Lane Changing Models. In: Transportation systems engineering anonymous, pp 15.1–15.12
6. Lee HK, Barlovic R, Schreckenberg M, Kim D (2004) Mechanical restriction versus human overreaction triggering congested traffic states. Phys Rev Lett 92:238702-1–238702-4

7. Habel L, Schreckenberg M (2014) Asymmetric lane change rules for a microscopic highway traffic model. In: 11th international conference on cellular automata for research and industry (ACRI), Krakow, pp 620–629
8. Ghaffari A, Khodayari A, Arvin S, Alimardani F (2012) Lane change trajectory model considering the driver effects based on MANFIS. Int J Automot Eng 2:261–275
9. Yoon J (2011) Path planning and sensor knowledge store for unmanned ground vehicles in urban area evaluated by multiple laders
10. El-Hajjaji A, Ouladsine M (2001) Modeling human vehicle driving by fuzzy logic for standardized ISO double lane change maneuver. In: 10th IEEE international workshop on robot and human interactive communication, pp 499–503
11. Nilsson J, Sjoberg J (2013) Strategic decision making for automated driving on two-lane, one way roads using model predictive control. In: IEEE intelligent vehicles symposium (IV), pp 1253–1258
12. Xu G, Liu L, Ou Y, Song Y (2012) Dynamic modeling of driver control strategy of lane-change behavior and trajectory planning for collision prediction. IEEE Trans Intell Transp Syst 13:1138–1155
13. Xu G, Liu L, Song Z, Ou Y (2011) Generating lane-change trajectories using the dynamic model of driving behavior. In: IEEE international conference on information and automation (ICIA), pp 464–469
14. El-Hajjaji A, Ouladsine M (2001) Modeling human vehicle driving by fuzzy logic for standardized ISO double lane change maneuver. In: 10th IEEE international workshop on robot and human interactive communication, pp 499–503
15. Engedy I, Horvath G (2009) Artificial neural network based mobile robot navigation. In IEEE international symposium on intelligent signal processing (WISP), pp 241–246
16. Doctor S, Venayagamoorthy GK (2004) Unmanned vehicle navigation using swarm intelligence. In: Proceedings of international conference on intelligent sensing and information processing, pp 249–253
17. Tomar SR, Verma S (2012) Safety of lane change maneuver through a priori prediction of trajectory using neural networks. Netw Protoc Algorithms 4:4–21
18. H. Jula, E. B. Kosmatopoulos and P. A. Ioannou, "Collision Avoidance Analysis for lane Changing and Merging', IEEE Transactions on Vehicular Technology, vol. 49, pp. 2295–2308, 2000.
19. Y. L. Chen and C. A. Wang, "Vehicle Safety Distance Warning System: A Novel Algorithm for Vehicle Safety Distance Calculating Between Moving Cars', in IEEE 65th Vehicular Technology Conference (VTC), 2007, pp. 2570–2574.
20. G. Feng, W. Wang, J. Feng, H. Tan and F. Li, "Modelling and Simulation for Safe Following Distance Based on Vehicle Braking Process', in IEEE 7th International Conference on e-Business Engineering (ICEBE), 2010, pp. 385–388.
21. Y. Wu, J. Xie, L. Du and Z. Hou, "Analysis on Traffic Safety Distance of Considering the Deceleration of the Current Vehicle', in Second International Conference on Intelligent Computation Technology and Automation (ICICTA), 2009, pp. 491–494.
22. Y. L. Chen, S. C. Wang and C. A. Wang, "Study on Vehicle Safety Distance Warning System', in IEEE International Conference on Industrial Technology (ICIT), 2008, pp. 1–6.
23. D. D. Salvucci and A. Liu, "The Time Course of a Lane Change: Driver Control and Eye-movement Behavior', Transportation Research Part F: Traffic Psychology and Behaviour, vol. 5, pp. 123–132, 6, 2002.

COLOMBO: Exploiting Vehicular Communications at Low Equipment Rates for Traffic Management Purposes

Daniel Krajzewicz, Andreas Leich, Robbin Blokpoel, Michela Milano and Thomas Stützle

Abstract While most standardized vehicular communication applications aim on increasing traffic safety, the exchange of messages between vehicles and the environment may be used for other purposes at no additional hardware costs as well. One possible area of such applications is traffic management. Traffic management requires data about the state of the road network before being able to predict or control traffic. The COLOMBO project, co-funded by the European Commission, examines the possibilities to use data gained via vehicular communications for traffic management purposes.

Keywords Vehicular communication · Traffic management · Simulation · Environmental issues

D. Krajzewicz (✉)
Institute of Transport Research, German Aerospace Center, Rutherfordstr. 2, 12489 Berlin, Germany
e-mail: Daniel.Krajzewicz@dlr.de

A. Leich
Institute of Transportation System, German Aerospace Center, Rutherfordstr. 2, 12489 Berlin, Germany
e-mail: Andreas.Leich@dlr.de

R. Blokpoel
Imtech Traffic & Infra, Amersfoort, The Netherlands
e-mail: Robbin.Blokpoel@imtech.com

M. Milano
Department of Computer Science and Engineering, Università di Bologna, Bologna, Italy
e-mail: Michel.Milano@unibo.it

T. Stützle
IRIDIA Lab, Université Libre de Bruxelles, Brussels, Belgium
e-mail: stuetzle@ulb.ac.be

T. Schulze et al. (eds.), *Advanced Microsystems for Automotive Applications 2015*, Lecture Notes in Mobility, DOI 10.1007/978-3-319-20855-8_10

1 Introduction

Vehicular communication is the ability of vehicles to exchange messages between each other and the infrastructure. Several scientific and technical efforts have been undertaken during the past years to make vehicular communications work. These range from investigations about the physics at the chosen communication frequencies, over the development of routing protocols to schedule and pass information in vehicular ad hoc networks (VANETs), simulative investigations of applications that use the exchanged information, performing field operational tests, up to the definition of according standards. First V2X-enabled vehicles are expected to be rolled out in 2015 [1].

The mentioned standards include the definition of a “basic set of applications” [2] that build upon vehicular communication. Most of these applications realize functions that aim at increasing traffic safety. Examples of such functions are assistance systems that warn the driver in case of an approaching emergency vehicle or a wrong-way driver. Only few applications are dedicated to other purposes, such as infotainment, insurance services, or fleet management services.

One of the major paradigms used by the standardized V2X communication is to enable the V2X-equipped vehicle to sense (“see”) its surroundings by obtaining so-called “Cooperative Awareness Messages” (CAMs) [3] from other V2X-equipped vehicles. These messages are sent periodically, most probably with an adaptive rate between 2 Hz for usual situations up to 10 Hz when being in situations that may be dangerous, e.g. the vehicle recognizes too high lateral accelerations. Notably, the equipment rate is the major factor for the performance of such applications. A V2X application will only work when at least two of such vehicles or one vehicle and an infrastructure road side unit (RSU) meet. A second barrier when deploying safety-related applications is their intrinsic need to work reliably as a failure may directly cost human lives.

But the vast amount of information sent via CAMs could be exploited for other purposes as well. E.g., it could be used by road side units (RSUs). Conventionally, road traffic management uses sensors mounted under the road surface or above the road to count passing vehicles and their speed. Such detectors rely on inductive loop, radar, or video sensors. They are either connected to a centralized traffic management center or locally to a near-by traffic light controller. Such sensors are known to be reliable and have a lifecycle of several years. Though, their installation and their maintenance costs are high, especially when taking into account that the replacement often requires to block the traffic at the site due to the need to open the road surface or to access installations above the road.

The possibility to use V2X technology instead promises to cover the complete area within the communication range at a price of only one road side unit if all vehicles are equipped. This will probably be never the case. Instead, a gradual increase during the next years can be assumed. According to estimate studies, until the year 2020, the number will stay low, under 20 % or less [4].

Therefore, the COLOMBO project [5, 6], co-funded by the European Commission, develops solutions for tactical traffic management that work at low penetration rates. Two traffic management topics are being targeted: traffic surveillance and road control via traffic lights. Besides, several proposals for other, new kinds of surveillance applications are presented by the project. The solutions are designed and implemented as software modules for being evaluated in simulations. While involving state-of-the-art open source simulation software packages for this purpose, the COLOMBO project extends them and returns these results as open source in many cases.

This article presents some of the project's theories and results. It is structured as following: The next chapter describes the traffic management solutions developed within COLOMBO so far and gives an outlook at following work. Afterwards, the used simulation suite and its extensions added by the project are presented. The article ends with a summary.

2 Traffic Management Solutions

The major goal of the project is to develop traffic management solutions for traffic surveillance and for controlling traffic via traffic lights. In the following subsections, some of the major achievements are presented.

2.1 Traffic Surveillance

Different approaches for determining the state on the roads using vehicular communication exist. The work performed in COLOMBO concentrated on the development of methods that deliver the information needed by the developed traffic lights system (see next section). These traffic light systems require information about the state of traffic in the arms of the intersection to be controlled, namely vehicle speeds, vehicle counts, spatial vehicle density, or vehicle queue lengths.

This information can be partially gained using nowadays sensors. However, their drawback is that they usually do not observe an area, but a cross section of the road or of a single lane. Thus, vehicle counts at one or many cross sections can be obtained well, while spatial information, such as density or queue lengths need to be determined indirectly via data extrapolation techniques.

The project delivered four different solutions for determining these measures. Two of them work by exchanging information between vehicles only (V2V), the other two ones by interpreting information obtained from vehicles at RSUs (V2X). All four solutions will be outlined in the following, more complete descriptions may be found in the project's publications.

Within the first solution, vehicles determine the per-road information by forming groups, first. This is done by exchanging information following a newly designed

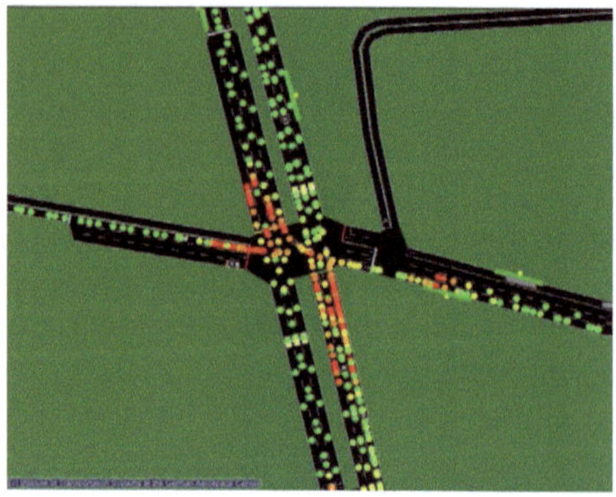

Fig. 1 Collecting CAMs at an intersection

protocol. Basically, the leader of the group is determined, first, and counts vehicles within his group or collects their speed information. The assignment of vehicles to roads is based on the information about the vehicle's driving direction, included in the exchanged messages. This approach is insofar interesting, as it allows to gain information about the road state without the involvement of static infrastructure units. The second algorithm determines the number of vehicles at a road section by measuring the delay of messages, what may be an option for extending the functionality of standard messages. Both algorithms resemble the "decentralized floating car data" (DFCD) metaphor: Single vehicles collect information and fuse it locally, instead of sending single messages to a central instance for aggregation like it is the case in conventional FCD systems.

Both remaining surveillance algorithms are assumed to be deployed at a road side unit. The first algorithm simply collects information given in the CAMs sent by vehicles that pass the RSU's communication range (see Fig. 1). The last one uses the travel time for computing the delays at an intersection, an often needed measure for controlling traffic lights.

The work in COLOMBO goes beyond investigations on methods for replacing conventional detectors. The usage of trajectories from V2X-enabled vehicles allows generating new views at intersections that can nowadays be only obtained using camera-based system. One of such solutions developed in COLOMBO is a local emissions monitoring system. Up to date, it is hardly possible to obtain such information in the real-world. Being a software application only, the system should be capable to be deployed at a road side unit at low costs.

The second novel approach for traffic monitoring pursued by the project is detecting traffic anomalies and incidents. It should be noted that determining accidents or critical driving maneuvers online is hardly possible at low equipment rates. This is due to the fact that the probability to see two equipped vehicles at once, additionally being involved or running towards being involved in an accident is very low. Still, possibilities to detect flow breakdowns exist and we see a high potential to achieve high detection rates of such anomalies at low V2X penetration rates. These methods are currently in the focus of COLOMBO work.

2.2 Traffic Light Control

Predominant traffic light control algorithms comprehend fixed time, green band and fully adaptive control algorithms. Within COLOMBO, the "SWARM" traffic light controller is being developed and tested: a self-organizing, adaptive, distributed and monitoring-aware approach to traffic light control that uses recent iterated racing optimization techniques (see next section). SWARM relies on a so-called macro policy: a concept for adaptive selection of so-called micro policies. A micro policy is a control algorithm for changing signal phases in a fixed or dynamic manner. After reviewing the state of the art and discussing with practitioners within the COLOMBO consortium, the following established micro policies have been taken into closer consideration.

- Static policy: the traffic lights loop continuously through the phases in a pre-defined fixed schedule. The program plans can be changed by a controller for activating pre-defined plans for the night or the weekend vs. rush hours, but this change usually occurs only few times a day. This approach is predictive: the logic is selected according to the expected amount of traffic, but does not take into account the real situation on the road or any unexpected variation in the flow of traffic.
- Actuated policy: traffic at all incoming lanes is measured using inductive loops [2]. As long as an incoming stream has green and a vehicle belonging to this stream passes the according detector, the actuated traffic light may extend the duration of the green light for this stream. The order of the phases is fixed as in a static policy.
- Phase policy: This policy will terminate the current chain as soon as another chain has reached the threshold, but respecting the minimum duration constraint of the current decisional phase. It is less reactive than the request policy, releasing the green light less often.
- Platoon policy: This policy tries to let all the vehicles in the currently green lanes pass the intersection before releasing the green light.
- Marching policy: This policy is adequate when the traffic looks too intense from all directions to take any online decision regarding the input lanes. In this case, there are two possible approaches: either use a static duration for decisional stages or consider the output lanes, do not allow traffic to lanes that are too heavily loaded.

The task of the macro policy is to select micro policies in respect to the current traffic state. The assumption is that every traffic state has one or more micro policies that are optimal. Therefore the strategy of the macro policy is to measure traffic state and decide which micro policy is best for this traffic state.

The measure of the traffic should be rather insensitive to very short peaks, like a singular platoon, but should react rapidly to more persistent traffic changes where we expect a burst in traffic from a single direction that will last for 15–20 min. For these reasons, an abstraction of the level of traffic is used, using simulated

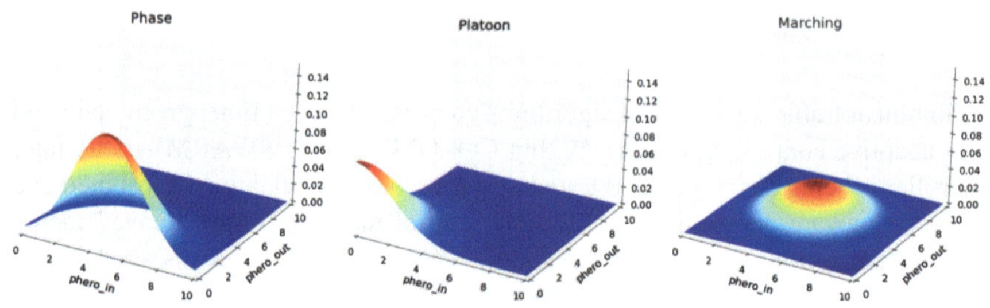

Fig. 2 Stimulus functions for different micro policies

pheromone levels. In nature, pheromone is an olfactory trail left by some animals like ants on the path they walk. This pheromone is additive: the more ants walk on a path, the higher the level of pheromone. Pheromone also evaporates over time, allowing ants use this as a guidance to choose their direction: the shorter the path from home to food, the less the pheromone evaporates and the higher the level.

Within COLOMBO, pheromone levels are used to calculate the level of traffic: cars driving down a lane or waiting at a red traffic light leave a virtual pheromone trail. It will quickly sum up if a significant increase of the traffic volume happens, and will evaporate in a short time when the number of cars decreases. The pheromone level at a junction is computed separately for incoming and outgoing lanes, but using the same equation structure. It accumulates the differences between the mean vehicles' speed and allowed speed [7]. The pheromone level parameters of incoming and outgoing traffic span a two dimensional parameter space. A probabilistic effectiveness score can be assigned to each micro policy for each point in this parameter space, resulting in some stimulus function (Fig. 2).

The macro policy of the swarm based traffic control algorithm investigated within COLOMBO tries to decide for the best micro policy based on some objective function. This is being achieved with the help of recent optimization techniques.

2.3 *Iterated Racing Based Optimization*

Traffic light control algorithms in general as well as the COLOMBO traffic light control algorithm incorporate numerous parameters (tuning parameters). They perform well in real situations when the corresponding tuning parameter values are chosen properly, raising demand for calibration procedures. As the number of parameters rise, manual calibration procedures become more complex. One may assume that the more sophisticated a traffic control algorithm gets, the more unlikely the best tuning parameter value set is to be found by even experienced practitioners. The SWARM traffic control algorithm developed in the COLOMBO project incorporates more than 50 tuning parameters. These tuning parameters

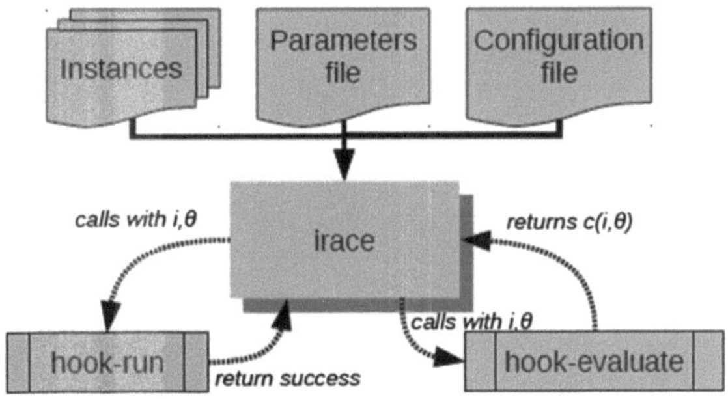

Fig. 3 Scheme of irace flow of information

address the traffic data acquisition process from incomplete V2X based measurements, tuning parameters of the micro policies and parameters for modeling the stimulus function.

Viewing the whole system as a black box, optimal parameter values can be found using so-called automatic algorithm configuration software (AAC software). Within COLOMBO, the iterated racing approach has been investigated in this context. The racing approach [8] is effective at identifying the best configuration from a given initial set of configurations. It proceeds by step-by-step testing the set of candidate configurations on new example instances of the problem being tackled (Fig. 3). At each step, it is checked by statistical testing whether differences exist among the candidate configurations and if it is so, then inferior candidate configurations are eliminated from further testing.

Tests were run on cluster computers using the free AAC software irace, simulating a synthetic crossroads scenario with SUMO (Fig. 4). Useful insight into inherent properties of the methodology has been gained regarding different research questions. The number of necessary iterations needed for achieving good results in parameter tuning has been investigated and it has been discovered that traffic control algorithm configurations that perform well at low V2X penetration rates are likely to perform well at higher penetration rates too.

The extensions of the irace AAC software performed in COLOMBO were released to the public. Additionally, a tuning tool kit "*tuningTK*" has been developed, tested and publicly released as free software. This tool allows a common interface to a set of AAC software packages.

2.4 Emissions Optimization

The reduction of emissions produced by road traffic is targeted in different parts of the project. The developed traffic light algorithms are designed in a way that fosters soft, environment-friendly modes of transport. Additionally, the project has delivered information about how to set up traffic light timings for reducing emissions for both, synchronized as well as sole traffic lights.

Fig. 4 Synthetic scenario for AAC software experiments

But emissions may as well be reduced by changing each vehicle's individual progress through the road network. This work topic is closely related to the GLOSA (green light optimal speed advisory) application, one of the very few V2X applications that belong to the "basic set of applications" and target traffic efficiency. GLOSA receives messages from road side units via vehicular communication (I2V). The messages contain the information about the current and future signal timings (within so-called SPAT messages) and the road infrastructure ahead (in so-called INFRA messages). Given this information, GLOSA can compute the speed to pass the next traffic light at green—saving energy and thereby reducing emissions by reducing the number of decelerations and accelerations. This speed is given as an advice to the driver.

Within the COLOMBO project, existing descriptions of the speed advice computation have been investigated. Additionally, own approach functions (see Fig. 5) that perform better than the existing ones in means of emissions have been developed. This work was based on determining real-world optimal acceleration and deceleration values as well as the optimal cruising speed.

3 Simulation System

COLOMBO evaluates the possibilities to use information from a low number of vehicles equipped with communication technology by designing according solutions and benchmarking those using simulations. The requirements for simulation systems that allow such investigations are well-covered in literature. Within COLOMBO, an established simulation system is used, originally designed

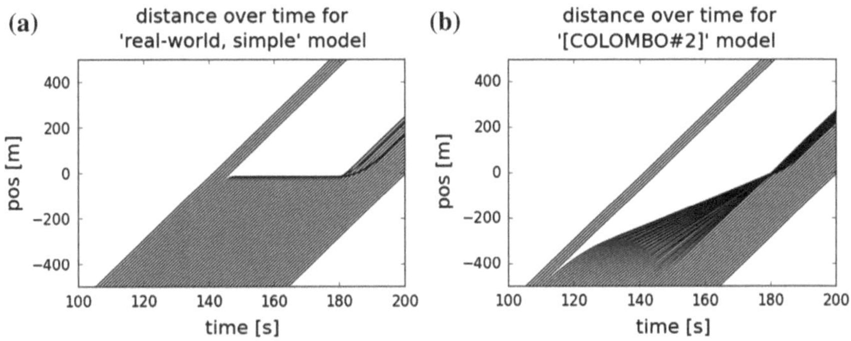

Fig. 5 Trajectories of vehicles approaching an intersection controlled by a traffic light; **a** assuming a simplified real-world behavior with no advice, **b** using the algorithm developed in COLOMBO

and implemented within the "iTETRIS" project. As outlined in the following, the system was extended by pollutant emission models and interfaces to the software configuration framework.

Several extensions have been performed on the simulation software's components. The following subsections describe the used simulation system and its components in larger detail, outlining the performed extensions.

3.1 COLOMBO's Simulation Architecture

The simulation system used within the COLOMBO project was initially developed within the "iTETRIS" project [9, 10] co funded by the European Commission between the years 2008 and 2011. It is a federating system built upon the well-established communication simulation ns-3 [11] and the traffic simulation SUMO [12, 13]. Both applications are joined using a middleware named "iCS", an abbreviation for "iTETRIS Control System" [9]. Besides both simulators, a simulation model of the application to evaluate is connected to the iCS. All connections are realized using sockets.

The simulation is performed step-wise. The iCS retrieves the positions of all vehicles from the road traffic simulation SUMO. If it sees a new vehicle, the iCS decides whether it should be equipped with a V2X device. For these vehicles, the iCS schedules messages to be simulated within the communication simulator ns-3. Respectively, it sends position updates for these vehicles to ns-3 during the subsequent simulation steps until the vehicle leaves the simulation. ns-3 informs the iCS about messages the V2X-enabled vehicles receive. This information is passed to the simulated V2X-application. The effects of the decisions the simulated application takes are resembled by sending according actions to the traffic simulator. SUMO's powerful socket-API "TraCI" allows to control a large set of simulated artefacts, including the simulated vehicles, traffic lights, detectors, etc.

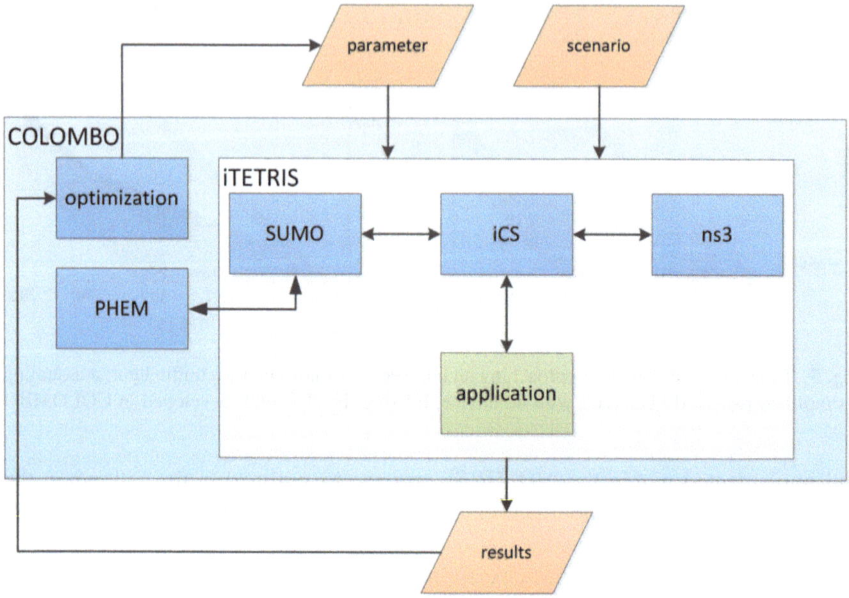

Fig. 6 The complete simulation system used in COLOMBO

For being applicable to the tasks scheduled in COLOMBO, the system has been extended by a new emission model and by interfaces to the online configuration tool, yielding in a system as shown in Fig. 6.

3.2 *Emissions Modelling and Optimization*

COLOMBO develops solutions that help in reducing the environmental impact of traffic. For a proper benchmarking of such solutions, a valid model of vehicular emissions is needed. Such a model should: Cover the complete vehicle fleet (in means of emission classes); Offer a classification of classes into Euro-norms; Compute certain pollutants (CO, CO_2, NO_x, PM_x, HC, and fuel consumption were chosen); (Be) sensible to microscopic parameters available in the simulation (mainly the vehicles' instantaneous acceleration and speed); Require only information that is available in the simulation; (Be) able to compute emissions in simulated time steps; (Be) easy to parameterize; (Be) portable, fast in execution, and directly embedded into the simulation [14].

The "PHEM" ("Passenger and Heavy Vehicles Emission Model", [15, 16]) model covers the complete vehicle fleet and has proved to be valid and sensitive to the factors named before. It is as well the base for some well-established national and European inventory emission models, such as HBEFA or COPERT. Still, being an application by itself, it is hardly applicable in iterative configuration runs, as required within COLOMBO. Therefore, a derivative of PHEM named PHEMlight was developed [14].

PHEMlight is directly embedded into SUMO. Each simulated vehicle may be assigned to a certain emission class via the vehicle's vehicle type. At each simulation step, the power the vehicle needs to achieve the wanted acceleration is computed. This power value is then normalized and used to look up the accordingly consumed amount of fuel and emitted pollutants. The data needed to compute and normalize the power as well as data needed to obtain the fuel consumption/pollutant emissions are loaded from additional data files.

PHEMlight's correctness has been shown by comparing its results against those obtained from the original PHEM model. This was done using an additionally implemented converter that parses vehicle trajectories generated by the traffic simulation and generates the according input for PHEM.

While the implementation of PHEMlight is available as open source, only two of the covered 112 vehicle classes are available for free. The complete data set may be purchased from the Institute for Internal Combustion Engines and Thermodynamics (IVT) at the Graz University of Technology [16], who is the major author of the model.

3.3 Pedestrians and Bicycle Modelling

The work on designing environmentally friendly solutions includes investigations on the prioritization of "soft" transport modes—pedestrians and bicycles. Although SUMO was capable to simulate multi- and inter-modal trip chains at the project's begin, no pedestrian dynamics were implemented, yet. When walking, pedestrians moved along the roads they had to pass with a constant speed and ignored any kinds of intersections by jumping over them. For COLOMBO, a finer representation of walking in urban environments was needed. The most important feature was to properly resemble the crossing of intersections controlled by a traffic light.

During the project, SUMO was accordingly extended. The road network now includes representations of "sidewalks", "crossings", and "walkingareas" (Fig. 7). These changes to the network introduce new elements only and are thereby backwards compatible to earlier versions of SUMO's network format. In accordance, the network importing application "netconvert" that comes with the SUMO simulation suite was extended by abilities to read given information about pedestrians and bicycle lanes from formats such as OpenStreetMap and by methods for determining this information heuristically in the case it is not given. As well, the routing modules had to be extended for reading and computing pedestrian routes through the given road network. Edges in SUMO's road network graph are unidirectional. As person may use a sidewalk in both directions, the routing computation had to be accordingly extended.

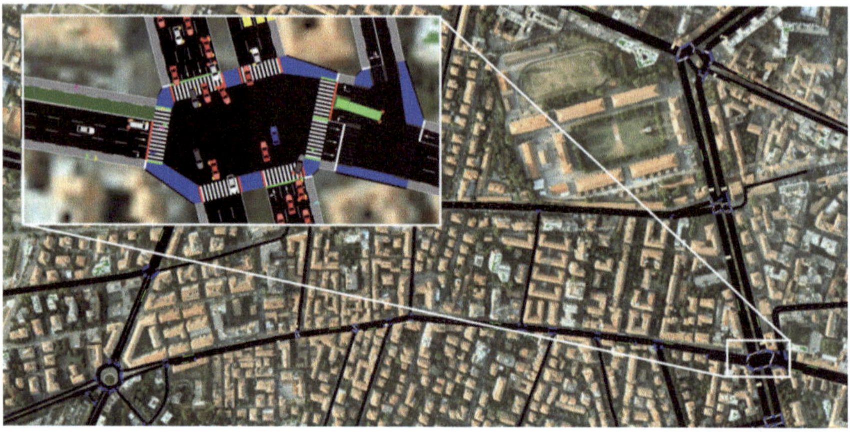

Fig. 7 A scenario with pedestrian enhancements. *Pedestrians* are shown at exaggerated size to increase visibility

The pedestrian dynamics themselves are "hidden" behind an interface. This allows to extend the simulation by new models for pedestrian dynamics. Currently, two pedestrian dynamics models are implemented. The first ("nonInteracting") resembles the simulation's earlier behavior, moving pedestrians with a constant speed along roads, disregarding any kind of interferences. The second model ("striped") is a new development. It is a one-and-half dimensional model where pedestrians move continuously ahead, but where their lateral behavior is modeled by "jumping" between "stripes" the available infrastructure is divided into. The simplicity of the pedestrian dynamics interface allows an easy inclusion of other models.

3.4 Scenarios and Methodologies

While a plethora of publications that report the use of traffic and communication simulations for evaluating the performance of V2X applications exists, one can hardly find well-defined and applicable road traffic scenarios one could rely on. This has been reported by different researchers already and was as well shown within the project. To overcome this issue, the COLOMBO project develops both, abstract synthetic scenarios, as well as ones that resemble parts of real-world road networks. The real-world scenarios are released continuously. The first ones that were made available resemble parts of the city of Bologna. A report on the released scenarios is [17].

4 Summary

The work performed within the first two years of the COLOMBO project was presented. The major task of the project is to exploit the possibilities to use vehicular communications for traffic management purposes assuming low rates of vehicles equipped with this technology. A motivation for this project and the developed solutions were described. The solutions mainly cover the areas of traffic surveillance and traffic light control. Additionally, the project's solutions for emissions reduction were outlined. COLOMBO develops the named solutions "in vitro" using simulations. The extensions performed on the simulation suite used within the project were outlined. The involved software packages are extended within the project.

Acknowledgements The authors want to thank the COLOMBO team for the great work performed so far. Additionally, we would like to thank the European Commission for co-funding the project under the grant number 318622.

References

1. Masters J (2012) Cars on the verge of talking. ITS International, November/December 2012
2. ETSI (2010) Intelligent Transport Systems (ITS); Vehicular Communications; Basic Set of Applications; Part 1: Functional Requirements. ETSI TS 102 637-1 V1.1.1 (2010-09)
3. ETSI (2014) Intelligent Transport Systems (ITS); Vehicular Communications; Basic Set of Applications; Part 2: Specification of Cooperative Awareness Basic Service. ETSI EN 302 637-2 V1.3.1 (2014-09)
4. Erdem B, Bansal P, Kühnel C, Geistefeldt J, Eschke S, Stahnke G, Grigutsch R, Schmidt J, Gizzi F, Jäkel C, Gläser S Deliverable D5.5, TP5-Abschlussbericht – Teil B-4, simTD, Page 83
5. Krajzewicz D, Heinrich M, Milano M, Bellavista P, Stützle T, Härri J, Spyropoulos T, Blokpoel R, Hausberger S, Fellendorf M (2013) COLOMBO: investigating the potential of V2X for traffic management purposes assuming low penetration rates. ITS Europe 2013, Dublin, 04–07 June 2013
6. COLOMBO consortium, COLOMBO web site, 2012–2015, http://colombo-fp7.eu/. Last visited on 31 Mar 2015
7. Hebenstreit C, Fellendorf M, Belletti R, Bonfietti A, Blokpoel R, Milano M, Niebel W, Krajzewicz D (2014) COLOMBO: deliverable 2.2 Policy Definition and dynamic Policy Selection Algorithms, 2014
8. Maron O, Moore AW (1997) The racing algorithm: model selection for lazy learners. Artif Intell Res 11(1–5):193–225
9. Rondinone M, Maneros J, Krajzewicz D, Bauza R, Cataldi P, Hrizi F, Gozalvez J, Kumar V, Röckl M, Lin L, Lazaro O, Leguay J, Haerri J, Vaz S, Lopez Y, Sepulcre M, Wetterwald M, Blokpoel R, Cartolano F (2013) ITETRIS: a modular simulation platform for the large scale evaluation of cooperative ITS applications. In: Simulation modelling practice and theory, Elsevier, doi:10.1016/j.simpat.2013.01.007, ISSN 1569-190X
10. iTETRIS consortium, iTETRIS web site, 2008–2015, http://www.ict-itetris.eu/. Last visited on 31 Mar 2015
11. ns-3 developer. ns-3 web site, https://www.nsnam.org/. Last visited on 31 Mar 2015

12. Krajzewicz D, Erdmann J, Behrisch M, Bieker L (2012) Recent development and applications of SUMO—simulation of urban mobility. Int J Adv Syst Meas 5(3&4):128–138. ISSN 1942-261x
13. German Aerospace Center (DLR). SUMO web site, https://sumo.dlr.de/. Last visited on 31 Mar 2015
14. Krajzewicz D, Hausberger S, Wagner P, Behrisch M, Krumnow M (2014) Second generation of pollutant emission models for SUMO", SUMO2014—second SUMO user conference, 15–16, May 2014, Berlin. ISSN 1866-721X
15. Hausberger S, Rexeis M, Zallinger M, Luz R(2009) Emission factors from the model PHEM for the HBEFA Version 3, Report Nr. I-20/2009 Haus-Em 33/08/679
16. Technical University of Graz: pages of the Institute for Internal Combustion Engines and Thermodynamics (IVT). http://portal.tugraz.at/portal/page/portal/TU_Graz/Einrichtungen/Institute/oe_123, 2005–2015. Last visited on 31 Mar 2015
17. Bieker L, Krajzewicz D, Morra AP, Michelacci C, Cartolano F (2014) Traffic simulation for all: a real world traffic scenario from the city of Bologna. SUMO 2014, 15–16 May 2014, Berlin

Optimal Traffic Control via Smartphone App Users

A Model for Actuator and Departure Optimisation

Daphne van Leeuwen, Rob van der Mei and Frank Ottenhof

Abstract For many years traffic control has been the task of traffic centres. Road congestion is reduced via traffic control based on the sensor information of the current traffic state. Actuators are used to create a better spread and throughput over the network. A powerful means to further reduce congestion is to shift from the classical reactive paradigm to a proactive paradigm. In this concept the traveller is included in the traffic control process in the sense that travellers are given advice about their travel scheme. This travel scheme presents the predicted travel time depending on time of departure and selected route. Today people use their smartphones to navigate. Via GPS and smart phone applications they optimise their route. Most of these applications use static traffic state information. In our research we develop a method to reduce congestion delay by including user decisions. According to the travel time preferences of the user a departure time/travel time curve is presented to the user. This curve shows the expected travel time corresponding to a specific departure time. Actuators are adapted according to the expected departure times of the app users. By including travellers information and preferences we want to analyse the resulting throughput and corresponding travel time in the network. To this end we study these effects for a small network with large peak arrivals in a short time period. Actuators in this network are adapted to the expected traffic flow and optimised accordingly.

Keywords Dynamic traffic management · Event control · Routing · Scheduling · Queueing theory

D. van Leeuwen (✉) · R. van der Mei
CWI, Science Park 123, 1098XG Amsterdam, Netherlands
e-mail: daphne.van.leeuwen@trafficlink.nl

R. van der Mei
e-mail: R.d.van.der.mei@cwi.nl

F. Ottenhof
Trinité Automation, J.N. Wagenaarweg 6, 1422, AK Uithoorn, Netherlands
e-mail: Frank.ottenhof@trinite.nl

T. Schulze et al. (eds.), *Advanced Microsystems for Automotive Applications 2015*,
Lecture Notes in Mobility, DOI 10.1007/978-3-319-20855-8_11

1 Introduction

Informing and routing travellers used to be primarily a task of the government. Due to the availability of real-time travel information this has been shifted to market parties. Nowadays numerous market initiatives have been developed to influence travellers via their own channels including travel apps, websites, navigation systems etc. These initiatives inform the traveller in route and departure time. Users can adapt their route and departure time choice according to this information. This concept should result in a more efficient use of the road network. Unfortunately this approach still lacks some important aspect. The user can only adapt his or her decision whereas the network actuators are taken as static tools. Traffic actuators only respond to current arriving traffic and do not adapt to decisions made by the user. By adjusting not only the users departure time and route choice, but also traffic actuators accordingly a tremendous decrease in delay could be established.

A concept is initiated to tackle this problem. This concept has been initiated by the name the Digital Road Authority. In this concept the road network is divided into small sub-areas and the road authority of each sub-area is coupled to a virtual road authority; the Digital Road Authority. This tool merges all traffic data into a smart travel advice system. It establishes a connection between public and private parties. Through a collaboration of the connected parties a more effective advice can be given. Via a smartphone application users can receive updates of the current traffic state relevant for them. This user, in his turn, sends information regarding route and departure time decisions to on-route traffic actuators. These traffic actuators collect information and adapt their setting accordingly. The Digital Road Authority plays a coordinating role between traveller and traffic actuators. The platforms in which this concept is developed consists of an unique collaboration between Dutch companies, knowledge institutes and the government (the triple helix). Thereby ensuring theoretical correctness and applicability.

2 Case Study: Traffic Control at Events

To test the Digital Road Authority in a real-life setting a case study will be used. For this case study it is essential to inform a significant amount of travellers in the studied area. A case study that satisfies these requirements would be during a large event. We introduce an area manager that coordinates the arena area. In this section this problem will be outlined.

During events large delays are often encountered for which not only the visitors of the event encounter delay, but also travellers with another destination. These events regularly cause problems in the nearby network, despite the fact that the number of visitors is known beforehand. These visitors depart their home uncoordinated and unaware of the choices made by other travellers. Traffic actuators are not adapted to the expected peak arrivals and therefore do not respond accordingly.

This is where the Digital Road Authority comes in. The Digital Road Authority coordinates, routes and informs travellers to facilitate the traveller to arrive at the event location with a minimal delay. Thereby collecting the user decisions and adapting the network to the expected peak arrival. This approach coordinates both travellers and actuators to facilitate an optimal throughput and minimal delay for the network around an event.

As a specific test case the area ArenaPoort in Amsterdam the Netherlands is used. This area is known for its event locations, Ziggo Dome, Arena, Heineken Music Hall. These locations attract many visitors. Problems arising during these events are:

- Parking problems, visitors driving around to search for available parking spots.
- Traffic jams due to long lines for parking lots.
- Large delays to exit the area after the event.

Via the Digital Road Authority a framework is created to inform coordinate and advice visitors and based on the visitor decisions controls traffic actuators. To accomplish this three phases can be distinguished.

Pre-event planning During this phase a plan is defined to give a personal departure and route advice to each traveller in a coordinated manner by using the available information. The available information that is used consists of three types. Via the ticket information the number of visitors and their area of departure is approximated. Secondly, the availability of parking lots and their capacity is known and last a map of the possible routes including their switch points are necessary. At a switch point a decision between two or more routes have to be chosen. A graphical representation can be seen in Fig. 1. Given these input sources an optimal arrival plan can be determined. The objective is to route visitors with minimal delay through the network by taking their preferred arrival time into account.

On-route adjustments As soon as the first travellers depart, adjustments can be made to the predefined plan. Given current road conditions it might be optimal to change parts of the route plan. Departed travellers are assumed to be fixed in departure time, hence they already left. They can however be rerouted or redirected to a different parking lot. Due to unpredicted events on the road or deviations from the original expectations the plan of the individual traveler can be optimised for the current state. An example of such a monitoring system is shown in Fig. 2. The occupation at the parking lots is shown, at the right from this the percentage of arrived visitors is monitored. Also the occupancy for each road is indicated by colour. Nearby visitors are monitored by counting these arrivals on each approaching road.

After-event The outflow after an event is easier to control by controlling the outflow pace from each parking lot. Given the current traffic state the optimal outflow stream can be determined in order to avoid congestion. Traffic actuators can be programmed in order to create nonconflicting routes to exit the event area to nearby highways.

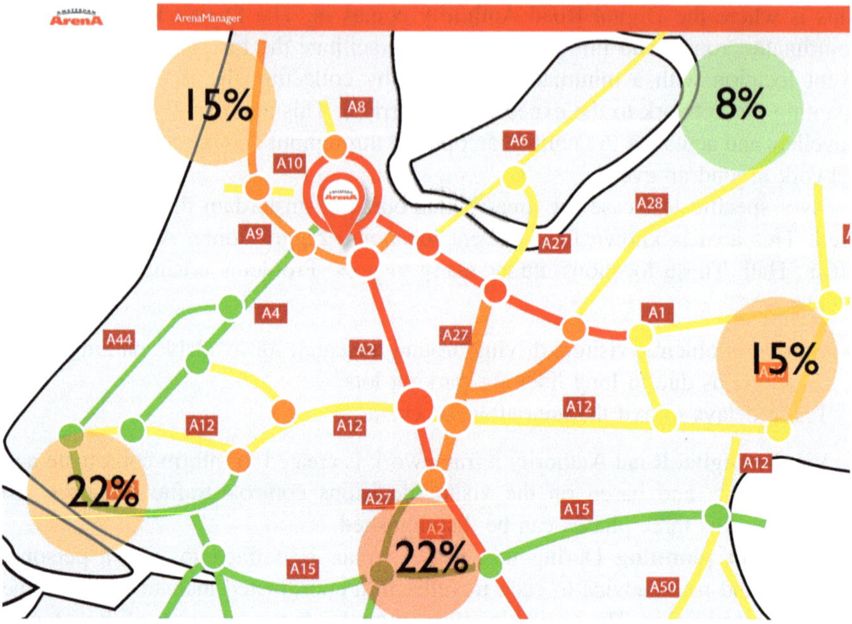

Fig. 1 A representation of the pre-event information regarding the origin of the event visitors and switch points

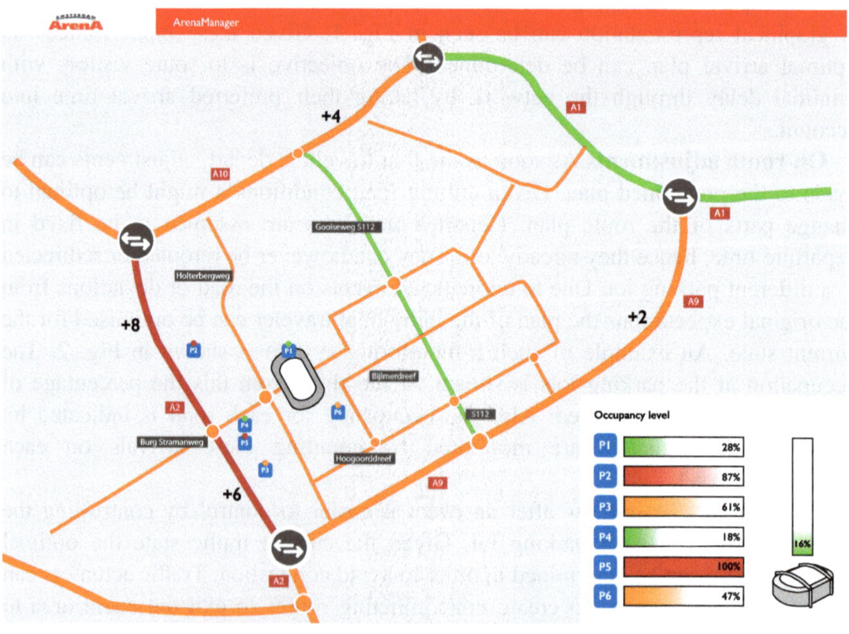

Fig. 2 A representation of on-route monitoring of the current state

3 Model Explanation

The developed model for this case study focusses on pre-event planning. The model will coordinate travellers via a managing agent defined as the Arena Poort area manager. Via a three fold optimisation model we can route the individual traveller and use its information to optimise traffic actuators for pre-event planning. The first part of the model maps the users to their destination without area manager interference, i.e. the current state of the traffic during events. Given these routes we can indicate whether unnecessary delays occur. If so, actuators can be controlled at switch points to redirect the traffic. These settings are used to see whether we would obtain a problem for the current number of expected visitors for an event. To improve the current situation we optimise in two directions. The actuators in the arena area are adapted to the expected arrivals due to the event and the visitors of the event are given a departure and route advice based on the users preferred arrival time. These steps influence each other and are optimised by iterating between both. Each step will be outlined in more detail.

Step 0: Initial setting

The origin of event visitors and the parking possibilities in the arena area are used to determine the expected route and delay. From each traveller the preferred arrival time is estimated. Based on this information the current delay during an event is measured. The delay obtained during this event is calculated.

Step 1: Actuator optimisation

In this model actuators along the routes to the arena area are adapted to improve the throughput in the arena area. Actuator settings are adapted according to the expected arrival stream of visitors before the event. These settings depend mostly on the throughput at parking lots per unit of time. A large amount of arrivals during a small time interval results in queues at the entrance of parking lots. Actuators have to respond to these expected queues by redirecting traffic to nearby parking lots. A queue before the entrance causes extra travel time for the visitor, if the queue exceeds a certain length this will cause spill backs to upstream roads. It is important to keep in mind that the adapted actuators not only influence the travel time of event visitors, but also the normal traffic through the area.

Step 2: Travel advice

A departure time is given to each user depending on the calculated routing possibilities, the amount of travellers and the parking capacities at the destination. The preferred arrival time, referred to as PAT, and departure location of the user are used as input parameters. Collecting this information from all users results in a large puzzle with numerous solutions. Via our model we want to find the optimal value of this puzzle. In other words, an optimal departure moment given the preference time of the user for which the overall delay is minimised. We can capture this in the following formula based on Vickrey's model [1]:

$$U(\tau) = \alpha D(\tau) + \beta ES(\tau) + \gamma LS(\tau),$$

where τ gives the expected arrival time, $D(\cdot)$ the deviation from preferred arrival time and $ES(\cdot)$, $LS(\cdot)$ give the penalties for early and late scheduled arrival times respectively. The parameters α, β, γ determine the penalty for a deviation from each of the parameters. The preferred arrival time deviations are calculated by:

$$ES(\tau) = \max(PAT - \tau, 0)$$

$$LS(\tau) = \max(\tau - PAT, 0)$$

The minimal value U for this formula has to be found. Each traveller wants to arrive as close as possible to their preferred arrival time combined with a minimal delay. Unfortunately, due to capacity limitations, people have to deviate from their preference during peak arrival periods. To model the expected delay over time for a predicted arrival pattern we obtain a waiting time curve over time, a visual representation can be seen in Fig. 3. Via this method the delay obtained by scheduling many users during the same time periods is captured. Via an optimisation algorithm using the above formula we can optimally spread the arrivals over time.

Step 3: *Stop condition*

To obtain the optimal setting multiple iterations of step 2 and 3 have to be performed. A change in departure time and route choice results in a changed delay function. This influences the performance of the actuator settings. We want to define the optimal actuator settings given the change in arrival pattern. Subsequentially the departure advice should be optimized according to the adaptation in the settings. The iteration steps will be performed until the difference between iterations is smaller than ε, for ε small.

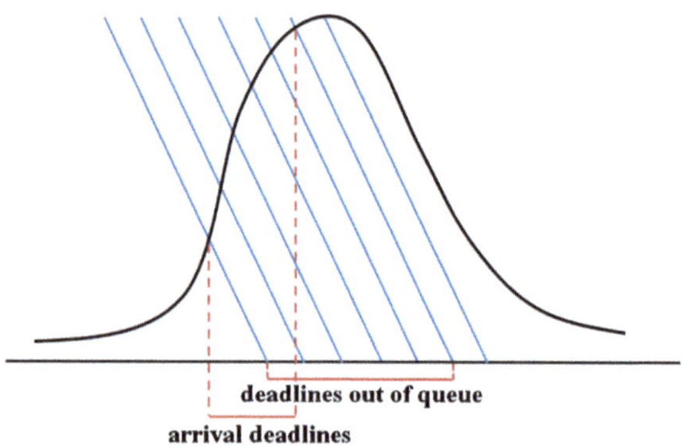

Fig. 3 Representation of the delay for expected arrival flow

4 Preliminary Results

From a previous case study we have shown that a travel advice application can result in a significant reduction of average delay. In this paper we use the same approach, thereby including the optimisation of traffic actuators in the optimisation process. We will outline results from a couple of scenarios in which we show the performance of the travel advice application.

In Fig. 4 results from two scenarios are shown. In this scenario travellers from one origin pass a road with fixed outflow capacity and a varying inflow over time. A queue results when the arrival stream is larger than the departure stream. By influencing part of the users to choose a deviation from their preference we can reduce the average delay over time. These users are rescheduled via the application of the Digital Road Authority project.

Results are shown in Fig. 4. The first scenario shows a small deviation of expected inflow over time. For various percentages of participating app users the decrease in delay is shown. At the second scenario a large increase of arrivals for a short time period is modelled. Also in this case a significant reduction in delay is obtained when we can influence 25 % of the total amount of travellers to change their departure time.

A second scenario considers two routes with fixed outflow capacity and one stream of travellers passing either one of the routes. The routes differ in length, the second one is longer. Therefore the second route will only benefits the user when the waiting time plus travel time of the first exceeds the travel and waiting time of the second.

In Fig. 5 results for the same scenarios are shown. In this case the user has the option to choose the other road. This strategy is very useful to incorporate for the arena case study. This area consists of many parking lots, which all have a varying distance to the event. Depending on the delay at the entrance of each parking lot it might be beneficial to choose a parking lot that requires some additional walking.

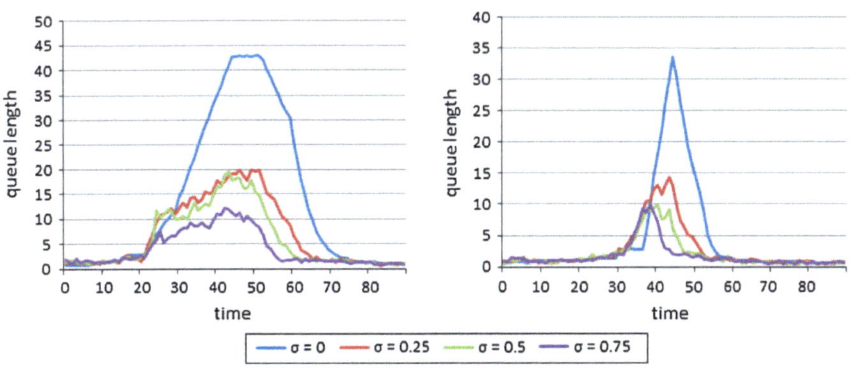

Fig. 4 Visualisation of the mean delay over time for two types of arrival scenarios for varying percentages of app user participants

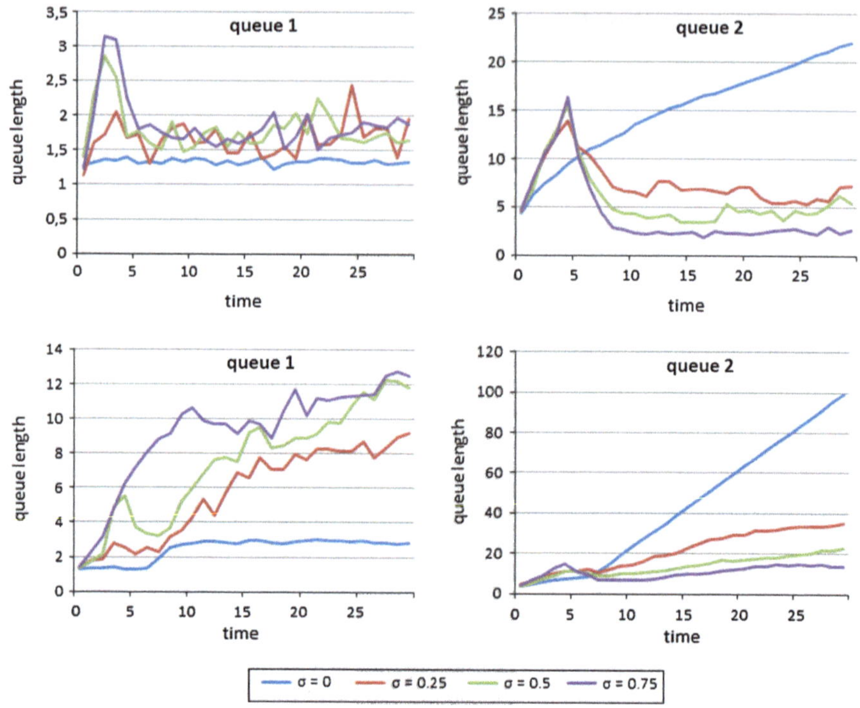

Fig. 5 Visualisation of the routing strategies over time for two arrival scenarios and varying percentages of app user participants

5 Conclusion and Further Research

This paper describes the model of an ongoing research project of the digital road authority. This case study focusses on the optimisation of the throughput in an area. Previously we focussed on establishing this goal via a travel advice, in this research we take the settings of the actuators into account in the optimisation process. Currently this research is still under development.

A detailed description of the model is visualised in Fig. 6. The arrivals are considered from four directions. At each input direction an actuator can define the split ratio of arrivals. The p-values along each road ending at the arena represent the sum of arrivals from the different directions. Depending on the parking lot capacity, arrival streams and the actuator possibilities a routing strategy is obtained. In the next step the travel advice algorithm improves the arrival streams over time.

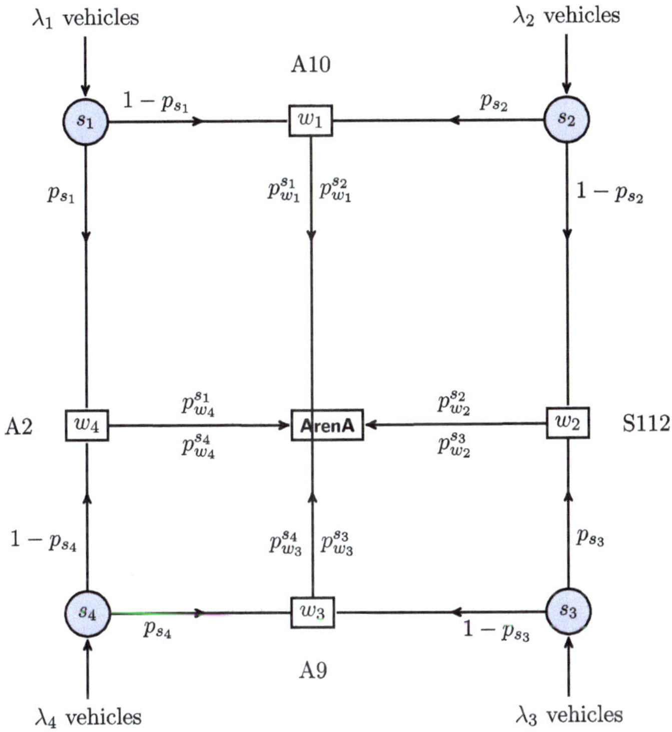

Fig. 6 Illustration of the arena area by the model parameters

Reference

1. Vickrey WS (1969) Congestion theory and transport investment. American Economic Review, Papers and Proceedings

Results of Mobile Traffic and Environmental Measurement for Green Traffic Management in the INTEGREEN Project

Reinhard Kloibhofer, Franco Fresolone and Roberto Cavaliere

Abstract The main objective of the EU LIFE+ project INTEGREEN is to introduce a demonstrative system for the municipal mobility management centre of the Italian city of Bolzano which can have the ability to provide distributed correlated traffic and air pollution information for the adoption of eco-friendly real-time traffic management policies. A new mobile probe with both traffic and air pollution monitoring units has been developed within the scope of the INTEGREEN project. The probe aims at overcoming the current limitation in the market of low cost measurement devices suitable for automotive use. The calibration of the air pollution monitoring unit has been performed in static conditions by comparison with official reference air quality stations, which can perform accurate measurements but at a low-speed. Several field measurement campaigns have been performed in the urban area of Bolzano and the most significant results are presented and evaluated.

Keywords Mobile air pollution measurement · Automotive application · Urban environmental traffic management

R. Kloibhofer (✉) · F. Fresolone
Digital Safety and Security Department, AIT Austrian Institute of Technology GmbH, 1220 Vienna, Austria
e-mail: reinhard.kloibhofer@ait.ac.at

F. Fresolone
e-mail: franco.fresolone@ait.ac.at

R. Cavaliere
TIS Innovation Park S.C.p.A., Via Siemens 19, 39100 Bolzano, Italy
e-mail: roberto.cavaliere@tis.bz.it

T. Schulze et al. (eds.), *Advanced Microsystems for Automotive Applications 2015*, Lecture Notes in Mobility, DOI 10.1007/978-3-319-20855-8_12

1 Introduction and Overview

In the EU LIFE+ project INTEGREEN the main objective has been the development of a demonstrative system for a municipal mobility management centre that can provide the public authorities (e.g. traffic operators and planners) with distributed and correlated traffic and air pollution information for the adoption of green and eco-friendly traffic management policies. The idea is that through a more detailed knowledge concerning both traffic and air pollution situation, authorities can more wisely adapt traffic and mobility strategies based on the measured conditions (e.g. the traffic lights' cycles on a specific route stretch), and learn from the field which specific policies can demonstrate to have in which conditions a significant impact in the reduction of air pollutant emissions and concentrations [1–3].

In order to enable such advanced scenarios a detailed knowledge of the traffic and air pollution conditions in a certain area in real-time is crucial. The technical solution proposed in the INTEGREEN project is a system that can continuously receive and process data gathered by driving vehicles as well as static roadside sensors.

Different research and innovation projects have been dealing with the final goal to better integrate traffic and air pollution dimension for more eco-friendly traffic management purposes, e.g. eCoMove (www.ecomove-project.eu) or CARBOTRAF (www.carbotraf.eu). Traditionally, the approach has been to use traffic data only to feed models of different complexity for calculating the current precise emission and dispersion of the air pollutants. The novel idea is to consider in such an architecture mobile air pollution measurements as well, which can be used e.g. to verify and/or calibrate the output of a dispersion model and eventually to provide richer information thanks to the higher resolution in space and in time.

Most of the air pollution monitoring units available at the state-of-art are designed for stationary applications (e.g. [4]). Only few and expensive examples, still at a research or in a pre-commercial deployment phase, offer solutions for handling the complexity of making sufficiently reliable measurements in mobility conditions. This situation represents therefore a significant barrier in the mass diffusion of such measurement systems and the large-scale testing of novel and integrated environmental traffic management strategies.

2 Environmental Parameters

For traffic-related applications different pollutants are of interest. One of the most important, in particular for the case study of Bolzano, is probably nitrogen dioxide (NO_2) and nitrogen monoxide (NO). Since with sun radiation NO_2 plus O_2 change to NO + O_3 (ozone), the latter pollutant can be taken into account in order to quantify the concentration of NO. In the INTEGREEN project, the final decision has been therefore to focus the attention to the following pollutants:

- Nitrogen dioxide NO_2
- Ozone O_3

Additionally, the most relevant meteorological parameters have been included in the monitoring process as well:

- Air temperature near the sensor
- Air relative humidity near the sensor.

3 Sensor Categories for Environmental Measurement

For the measurement of pollution gases like NO_2 and O_3, three main physical measurement principles for sensors can be distinguished:

3.1 Optical Gas Sensors

The basis of optical gas sensors is the Beer–Lambert law. It relates the absorption of light to the properties of the material through which the light travels.

The light absorption of the light travelling through a homogenous medium is described by the Beer–Lambert law:

$$I(z) = I_0 \cdot e^{-\alpha \cdot z}$$

where:

$I(z)$ … light intensity at the position z [W m^{-2}]
I_0 … light intensity at the position $z = 0$ [W m^{-2}]
α … absorption coefficient [m^{-1}]
z … distance [m]

The absorption coefficient depends on the material (gas), the wave length of the light, the air pressure and the temperature. For each gas exists a special wavelength where the absorption rate is a maximum. To measure the gas concentration of a specific gas a light source generates the light with the characteristic wave length of this gas. After the light travels through the measurement gas a photo detector on the other end of the tube measures the light intensity and calculates the absorption rate and the gas concentration (Fig. 1).

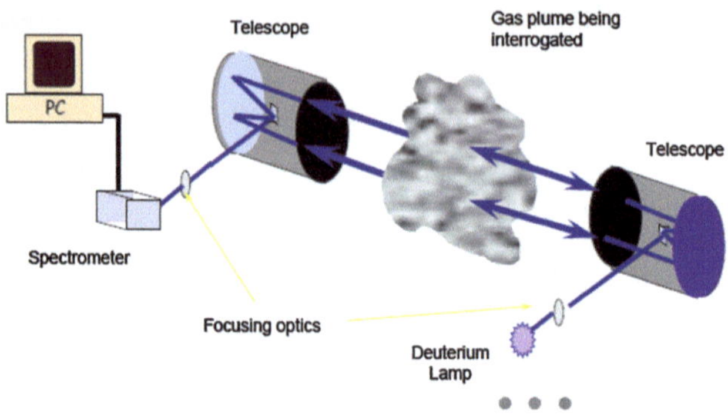

Fig. 1 Optical gas sensor

3.2 Electrometric Sensors

An electric potential is generated with an electrometric gas sensor as a function of the gas concentration. This potential can be measured with electronic devices, e.g. an analog to digital converter.

These electrometric sensors have three or four electrodes (Fig. 2). The working electrode (also called sensing electrode) is designed to optimise the oxidation or reduction of the measurement gas. To operate such a sensor a potentiostatic circuit is necessary to hold the electrical potential of the working electrode at a fixed value with respect to the reference electrode. By keeping the electrical potential fixed, a current from the working electrode flows in the potentiostatic circuit and generates an electrical output signal. This can be measured from an ADC.

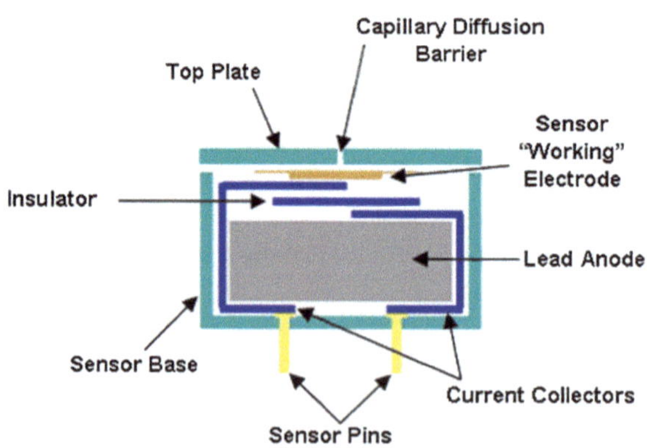

Fig. 2 Electrometric sensor

3.3 Chemo Resistive Sensors, MOX (Metal Oxide Semiconductors)

The structure of a Metal Oxide (MOX) sensor is shown in Fig. 3.

On a substrate with 2 metal electrodes an active material (i.e. SnO_2) is deposed. With the concentration of a specific gas the conductivity of this active material change. On the two metal electrodes the conductivity can be measured. An operation temperature of typically 300–400 °C below the substrate is necessary. This can be created with an electrical resistive heating (i.e. platinum) above the sensor.

The working principle is based on the variation of conductivity in the presence of oxidizing and reducing gases. The electron density changes with the adsorption and desorption of oxygen (O, O_2). Adsorbed oxygen gives rise to potential barriers at grain boundaries and thus increases the resistance of the sensor surface. The variable electric resistance is measured indirectly by providing a reference voltage on a voltage divider where one of the resistors is fixed while the other is the resistor of the sensor varying with the gas level. By measuring the signal on the varying resistor the gas concentration can be calculated.

All of these three types of sensors are very different in costs, mechanical properties, response time to gas level change, accuracy, stability and dependency in temperature and humidity [5]. The optical sensors are the most accurate sensors but up to now they are very expensive and large in size. Additionally the measurement value is averaged for a relative long time (e.g. 30 min) to further reduce measurement noise. Therefore they cannot be implemented in a small and low cost mobile monitoring probe. The MOX and the electrometric sensors have been therefore both tested in the new developed monitoring probe.

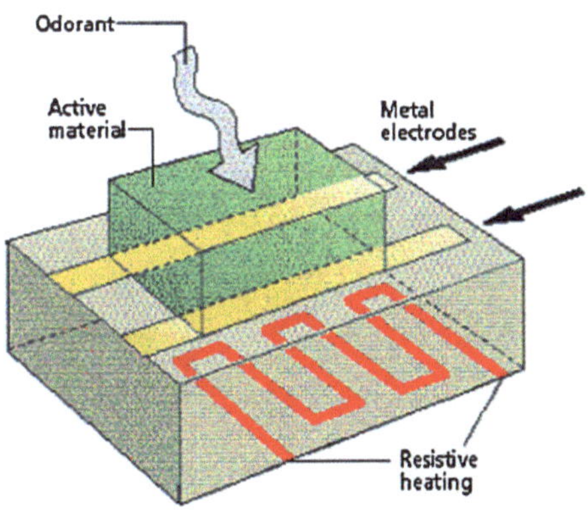

Fig. 3 MOX sensor

4 Forced Air-Flow

For a fast and mobile monitoring of gas values it is important that the probe air is in close contact with the sensors. If the natural air movement is very low the sensor measures always the same value. An additional problem for the MOX sensor is the variation of air flow speed on the sensor surface because the sensor is heated to a high temperature and if the air flow changes the sensor temperature and consequently the sensitivity of the sensor will also change.

To overcome these problems an ad hoc air guide above the sensors has been designed (see Fig. 4). To generate a constant air flow on the sensors an air pump has been placed near the air guide and connected with a short piece of tube. The air flow is measured with a dedicated air-flow sensor to monitor and keep constant the flow.

The compact measurement sensor platform is illustrated in Fig. 5. The connection of the air input and air output can be seen on the right side of the probe. For aspiration from a dedicated area outside the vehicle a flexible air tube is connected on the air-in connector of the sensor platform.

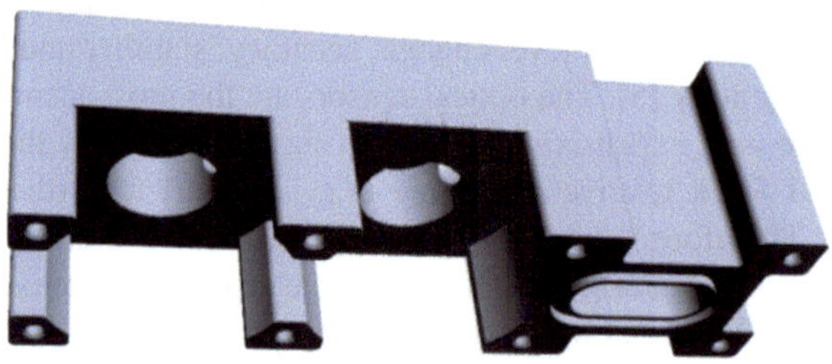

Fig. 4 Ad-hoc air guide final design

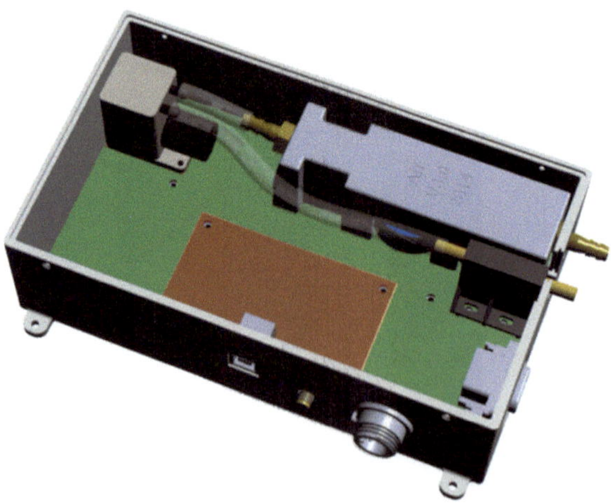

Fig. 5 Mobile air pollution monitoring unit final design

5 Measurement Probe

To build a portable monitoring probe with automatic upload of the data additional components are necessary.

The raw data of the environmental monitoring unit is transmitted to a telematics unit where the data can be temporary stored and pre-elaborated. A Windows based platform was selected because the SW for pre-elaboration is simple and fast to develop with tools developed on such an operating system. On the other hand, the system must be suitable for mobile and automotive purpose. So an embedded platform with 12 V/24 V automotive voltage and with no moving parts inside like ventilator or hard disk was selected.

When the data from the air pollution monitoring unit is delivered to the telematics unit a time stamp and a geo-position to each data package is added. The geo-position is received from a GPS-receiver. To distinguish from more than one mobile platform a car-id number is added to the data package too.

All the data are stored in a local data base for later offline elaboration. But before writing the data to the data-base the row values must be pre-elaborated to get calibrated air pollution values. With a display connected to the telematics unit the real-time measurement results can be observed and with a special SW-tool also the historical data can be displayed in a graphical view.

With an integrated 3G/4G modem the monitored data is uploaded in blocks to a server platform. In the case that the modem has no connection to the Internet due to poor signal coverage, the data is not lost but transmitted at the next possible connection of the mobile device. To get a compact device the external antenna for the modem can be directly connected on the front side of the mobile platform.

The complete mobile platform is shown in Fig. 6. On the top side, the embedded telematics platform can be seen, while beneath up to two sensor platforms can be built in. At the right side there is the air pollution monitoring module and on the other side a traffic monitoring module with the possibility of connection to the vehicle CAN-bus.

Fig. 6 Complete INTEGREEN mobile monitoring probe

6 Calibration and Verification

One of the main challenges for such a measurement system is the calibration of the air pollution sensors. As mentioned before, low cost, fast and accurate reference measurement devices are currently hardly available. Consequently the calibration and evaluation of the measurement speed has been carried out in two different steps.

6.1 Calibration of Pollution Level

For an accurate calibration of the measurement device the mobile system was placed near a fixed air quality station (see Fig. 7) and the measurement data was recorded for a reasonable long time (e.g. half a day) in order to have sufficient parallel measurements.

The official air pollution measurements [6] have been compared with the data measured by the mobile monitoring unit. Calibration values have been calculated to have a minimum square error between both measurement systems. The comparison of the data acquired by the fixed air quality station and the corrected mobile measurement data monitoring unit is illustrated in Fig. 8.

Fig. 7 Calibration session of the INTEGREEN mobile air pollution measurement system

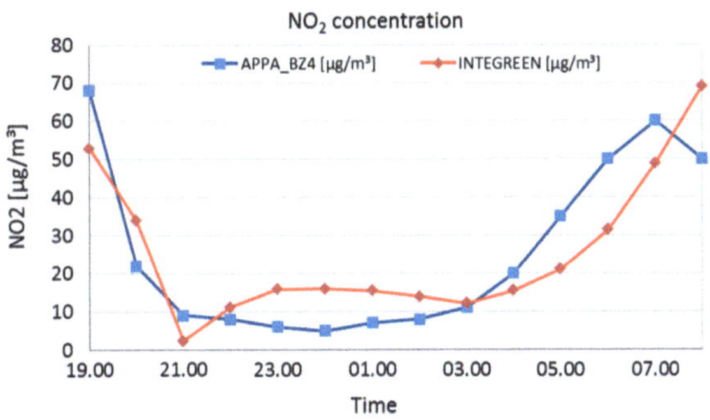

Fig. 8 Comparison of air pollution measurements: fixed air quality reference station (*in blue*); INTEGREEN mobile air pollution measurement system (*in red*)

It is important to underline that signal-to-noise ratio of calibrated measurements is further improved at server side through proper filtering techniques, e.g. moving averages, exponential and Kalman filters [7].

6.2 Evaluation of Measurement Speed

The evaluation of the response answer of the sensors on a sudden change of pollution level without a fast and accurate measurement device has been a big challenge. One of the open questions that the INTEGREEN project wanted to clarify through the mobile system has been in particular to understand the spatial variations of air pollution levels while driving on a road with medium speed (in urban area we assume 50 km/h).

To evaluate the ability of our measurement device in correspondence of sudden air pollution levels changes specific field test sessions were organized. In particular, a test car with the mobile measurement system on board has been driven through two consecutive tunnels. The drive has been repeated in both directions in order to verify similar response patterns.

One of the results associated to one of this test session is shown in Fig. 9. When entering the first tunnel, we observe a slow increase and the peak of pollution level reached in correspondence of the exit gate. While out of the first tunnel the pollution level decreases very fast in the space of some tens of meters, reaching again typical open-air pollution levels. When entering the second tunnel, we observe on the contrary a much stronger and sudden increase of pollution level, characterised by an approximately linear trend. In this case, the peak is reached before the middle of the tunnel, followed by a slow decrease of the pollution level up to the exit of this tunnel. At this point the direction of travel has been inverted, the same two tunnels have been driven but in the opposite direction. The measured concentration shows a specular pattern which fits very well with the one observed just some couple of seconds before.

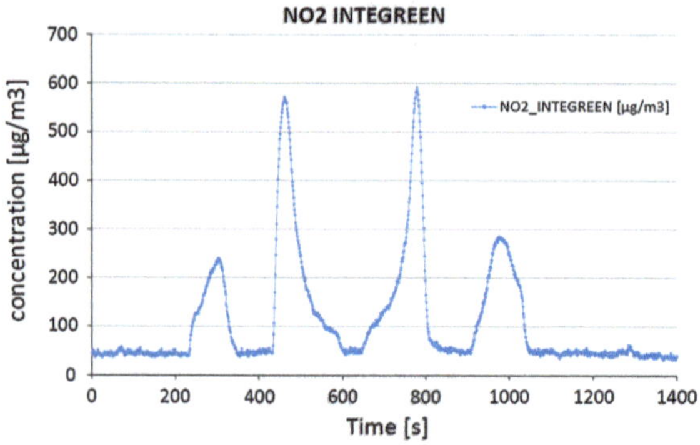

Fig. 9 Air pollution concentration in consecutive tunnel environments

These results demonstrate the ability of the developed measurement probe to be a good compromise in terms of measurement speed and accuracy. The empirical experience in INTEGREEN has suggested that the main contribution of such a system has to be intended more in qualitative terms than in quantitative terms: from an operational point of view, it is more important to know the level rather than the exact amount of air pollution concentration on a certain road stretch, and in particular if hotspot situations are present. On open roads the spatial variation of pollution levels is typical slower, but peaks in the order of 100 $\mu g/m^3$ can be observed.

7 Field Measurement and Results

To evaluate the mobile environmental probe in real urban traffic conditions, different field measurement campaigns have been performed. In Fig. 10 the pollution levels on the streets associated to a long test drive carried out in February 2014 is illustrated on a map. This driving session has been performed in the urban area of Bolzano and covers different areas like industrial zones, residential districts as well as the A22 highway crossing the city.

It can be observed some green tracks with lower pollution. Some red tracks inside the city have higher pollution levels caused by a higher traffic load with some limited situations of stop-and-go traffic. Higher pollution levels have been measured on some tracks on the highway but not on the whole of it. A typical increase happens if some trucks or busses drive uphill. It can also be observed, that on some crossing points the pollution level is increased (e.g. in the middle of the picture on the upper part). This is typical when the vehicles first are waiting at the red traffic light and then accelerate at the green traffic light.

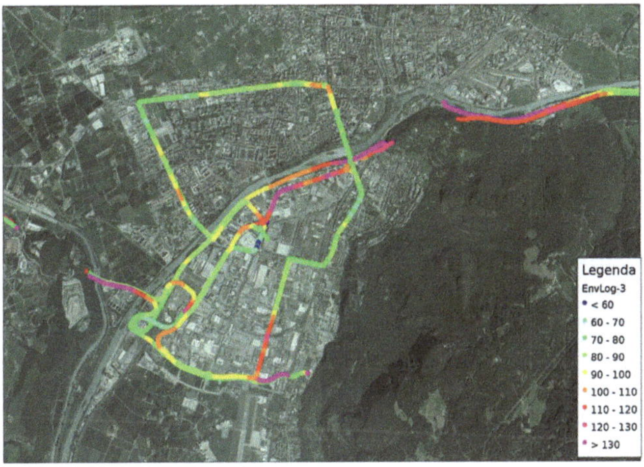

Fig. 10 Field measurement results (map presentation)

In Fig. 11 the results of another long measurement drive in May 2014 of about 50 min in Bolzano are presented. In this case NO_2 concentrations are illustrated over a time axis. It can be seen that in this case most of the time the pollution levels are between 35 and 45 μg/m^3 which is the common situation when meteorological conditions are favorable and traffic load as well as other pollutant sources contributions are limited. A higher peak near 500 s has been measured in a very short tunnel, the second peak at 2500 s is located near a crossroad at the city entrance where many vehicles are stopping and accelerating after a traffic light. The smaller peaks after 2500 s are due to stop-and-go traffic in correspondence to the center of the city.

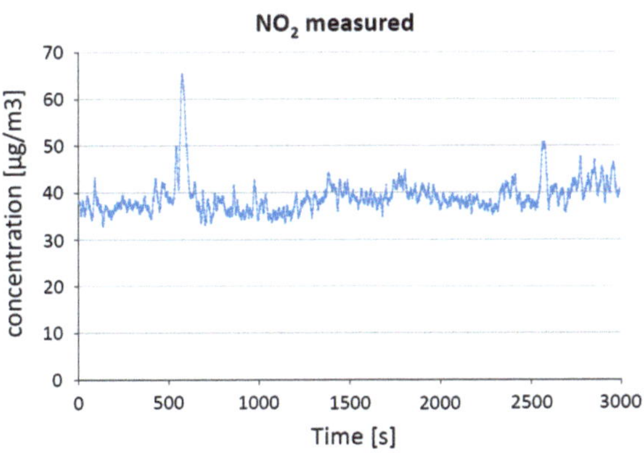

Fig. 11 Field measurement results (plot presentation)

8 Summary and Further Development

In the INTEGREEN project, supported by EU LIFE+ program of the European Commission, a demonstrative monitoring platform for the traffic management centre of the city of Bolzano has been developed. This platform is capable to monitor on a real-time basis traffic conditions as well as the associated environmental situation at the roadside level. One of the components of this complex system is a novel mobile monitoring system, which can provide traffic and air pollution measurements from on-board vehicles. Because low-cost and fast environmental measurement devices are at present not available on the market, a new system has been designed and developed from scratch.

First, an investigation of different sensor technologies and characteristics for NO_2 and O_3 sensors—among the most important pollutants as far as the traffic source contribution is concerned—has been performed. In order to have a fast and reliable sensor response at all driving conditions, an ad hoc air guide system with air pump and air-flow monitoring has been introduced. To properly read out the analog sensors, all necessary electronic was designed and developed as well. A telematic unit able to save and pre-elaborate measurement data and geo-information values with the capability to send the data in real-time to a host server has been selected and the related SW developed.

The verification and calibration of the mobile monitoring system prototype has been performed in two steps:

- Static calibration with reference air quality station
- Verification of response time through specific driving sessions in harsh environments (i.e. tunnels).

An extended field test campaign has been performed in the city of Bolzano and surrounding areas in order on one side to improve the local understanding on air roadside air pollution patterns, and on the other side to check how such a system can be used in an operational mode. The INTEGREEN central platform is capable on a real-time basis to present processed data in a geo-referenced way (i.e. on a map) as well on time plots.

This Mobile Probe developed and tested in the INTEGREEN project is planned to be extended with further sensors in a follow-up project and/or can be directly industrialised with interested stake holder.

References

1. Calcolo e valutazione delle emissioni di CO_2 e definizione di scenari di riduzione per la città di Bolzano, a study produced by EURAC research for the Municipality of Bolzano. http://www.comune.bolzano.it/UploadDocs/7528_BolzanoCO2_Report_Ita_100201.pdf, 2010
2. Valutazione dell'impatto sulla qualità dell'aria nella città di Bolzano, a study produced by CISMA for the Municipality of Bolzano. http://www.comune.bolzano.it/UploadDocs/9152_Qualita_aria2_27_01_2011.pdf, 2011

3. Comune Bolzano, Assessorato Mobilità, Piano Urbano della Mobilità 2020, Dicembre 2009
4. Provincia Autonoma di Bolzano - Agenzia Provinciale per l'Ambiente, Valutazione della qualità dell'aria 2005–2015 - Allegato 2, Bolzano, 2010
5. Aeroqual, [Online]. Available: http://www.aeroqual.com/wp-content/uploads/2014/01/AQM60-Brochure.pdf
6. Provincia di Bolzano - Agenzia Provinciale per l'Ambiente, Situazione dell'aria in Provincia di Bolzano, [Online]. Available: http://www.provinz.bz.it/umweltagentur/2908/luftsituation/index_i.asp
7. http://infoscience.epfl.ch/record/175158/files/mdm2012_final.pdf

Part IV
Electrified Power Trains and Vehicle Efficiency

Design and Fabrication of a SiC-Based Power Module with Double-Sided Cooling for Automotive Applications

Klas Brinkfeldt, Jonas Ottosson, Klaus Neumaier, Olaf Zschieschang, Eberhard Kaulfersch, Michael Edwards, Alexander Otto and Dag Andersson

Abstract The electrification of drive trains combined with special requirements of the automotive and heavy construction equipment applications drives the development of small, highly integrated and reliable power inverters. To minimize the volume and increase the reliability of the power switching devices a module consisting of SiC devices with double sided cooling capability has been developed. There are several benefits related to cooling the power devices on both sides. The major improvement is the ability to increase the power density, and thereby reduce the number of active switching devices required which in turn reduces costs. Other expected benefits of more efficient cooling are reductions in volume and mass per power ratio. Alternatively, improved reliability margins due to lower temperature

K. Brinkfeldt (✉) · D. Andersson
Swerea IVF, Argongatan 30, 431 53 Mölndal, Sweden
e-mail: klas.brinkfeldt@swerea.se

D. Andersson
e-mail: dag.andersson@swerea.se

J. Ottosson
Volvo Group Truck Technology, Sven Hultins Gata 9D, 412 88 Göteborg, Sweden
e-mail: jonas.jo.ottosson@volvo.com

K. Neumaier · O. Zschieschang
Fairchild Semiconductor GmbH, Einsteinring 28, 85609 Aschheim, Germany
e-mail: klaus.neumaier@fairchildsemi.com

O. Zschieschang
e-mail: olaf.zschieschang@fairchildsemi.com

E. Kaulfersch
Berliner Nanotest und Design GmbH, Volmerstr. 9B, 12489 Berlin, Germany
e-mail: eberhard.kaulfersch@enas.fraunhofer.de

M. Edwards
Chalmers University of Technology, 412 96 Göteborg, Sweden
e-mail: micedw@chalmers.se

A. Otto
Fraunhofer ENAS, Technologie-Campus 3, 09126 Chemnitz, Germany
e-mail: alexander.otto@enas.fraunhofer.de

T. Schulze et al. (eds.), *Advanced Microsystems for Automotive Applications 2015*,
Lecture Notes in Mobility, DOI 10.1007/978-3-319-20855-8_13

swings during operation are can be expected. Removing the wire bonds on the top side of the devices is expected to improve the reliability regardless, since wire bonds are known to be one of the main limitations in power switching devices. In addition, it is possible to design the package with substantially lower inductance, which can allow faster switching of the devices. In this paper the design, simulations and fabrication process of a double sided SiC-based power module are presented.

Keywords Power electronics · Silicon carbide · Double-sided cooling · Packaging · FE-simulation

1 Introduction

An integral part of the technology in electrification of the drive train in automotive applications besides the energy storage system and the traction unit is the power inverter module. The function of the power inverter is to transform a high DC voltage into current pulses required for the electric motor.

For all types of drive train architectures but especially in the case of distributed drive systems, including in-wheel or near wheel architectures, the drive train is separated into several units, where the power modules need to be highly integrated and compact [1, 2].

To base the power inverter on silicon carbide (SiC) power devices can reduce the size compared to using silicon-based counterparts. This is due to improved current density and thermal performance. Other benefits of SiC power devices include higher thermal conductivity, lower switching losses, and a capability to operate at higher switching frequencies and temperatures [3, 4].

Even though SiC devices are capable of operation in higher temperatures it is still important to lower operational and cycling temperatures as this leads to an increase of the life time and enlarged reliability margins of both the devices and their packaging. It is therefore essential to remove dissipated heat effectively from the switching devices regardless of the semiconductor material. Thus, improvements to the thermal management remain an important issue.

One way to reduce the thermal resistance is to achieve larger heat exchange area by cooling both sides of the power switching device chips. This promotes an over-all lower on-state voltage and an increase in the current carrying capability [5–7]. In addition, removal of the wire-bond connections at the top side contacts allows packaging designs, which have demonstrated to significantly reduce the switching cell inductance, which allows for high speed switching and even lower switching losses [8]. Furthermore, failure modes associated with the bond wires, such as lift off and heel cracking are removed.

In this work, the design and fabrication of a power module with double sided cooling is described. The power devices are sandwiched between two Direct

Bonded Copper (DBC) substrates and a new way to preserve the edge passivation stand-off distance by machining of the DBC copper layer [9]. Thermo-mechanical simulations comparing single sided cooling with the double sided cooling modules are also presented.

2 Design

A common power circuit for electric motors consists of a three-phase inverter bridge configuration. To reduce the parasitic inductance in the power stage and increase the modularity in an inverter design it makes sense to realize the circuit in the form of individual power modules each containing a half-bridge configuration as shown in Fig. 1. Three such half-bridges (phases) then form the three-phase circuit (Fig. 2). This modular concept increases the flexibility and reparability/ maintainability in the inverter housing design stage as well as the scalability, as more modules can be used in parallel to handle higher current loads.

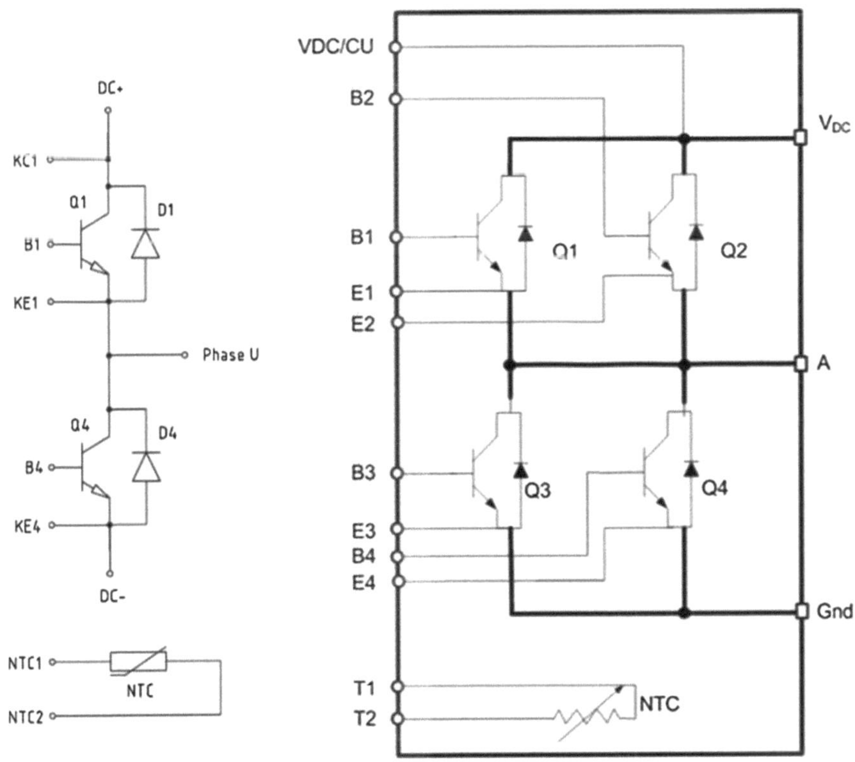

Fig. 1 Half-bridge configuration (*left*), and the content of a single module (*right*)

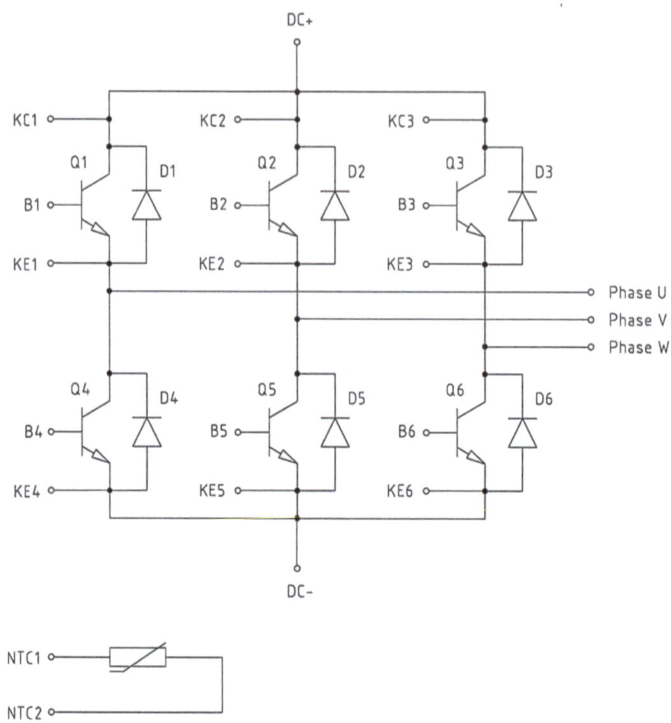

Fig. 2 Inverter circuit, three-phase bridge configuration

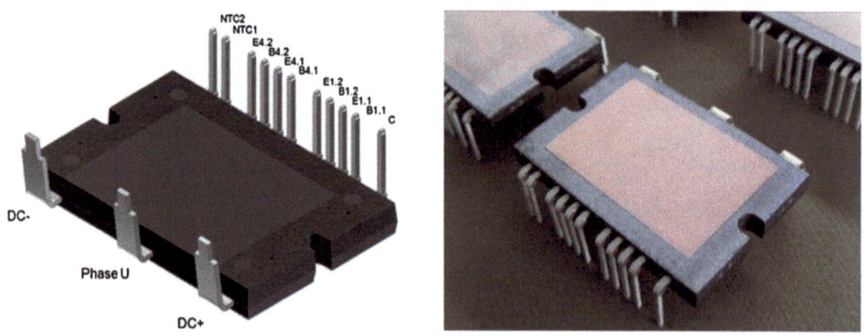

Fig. 3 The APM19 power module package with single sided cooling

For initial tests and for comparison reasons the package design of the power module is based on the existing single sided cooling APM19 module package (Fig. 3). This is an automotive qualified module package designed and manufactured at Fairchild Semiconductor. The size of the module is sufficient to accommodate four SiC bipolar junction transistors (BJTs) and accompanying four SiC diodes for each switch of the half-bridge.

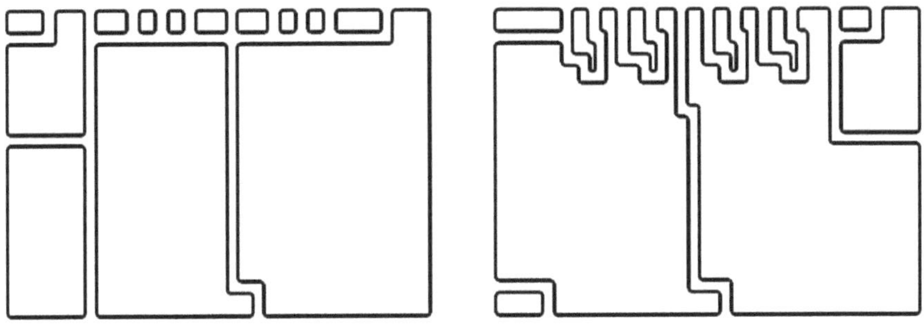

Fig. 4 Layout of the top and bottom copper layers of the DBC substrates

The DBC layout for the double sided module is designed to maximize the copper area for the current carrying paths while maintaining equal lengths for all base signals. The layouts of the top and bottom DBC substrates are shown in Fig. 4.

To connect the circuit a specially designed lead frame is used. In the double sided package it is important that the devices and lead frame share roughly the same thickness. The thicknesses of the SiC BJTs and diodes are 375 ± 10 μm and 387 ± 10 μm respectively. The lead frame thickness is required to be larger than that to be able to carry the amount of current expected. A thickness of 0.8 mm is used on the APM19 single sided version. Hence, the lead frame needed to be thinned down from 0.8 mm to approximately 375 μm at the parts which are sandwiched between the DBC substrates. The lead frame design is shown in Fig. 5.

The original metallization on the top side of the devices is aluminum to facilitate connection of aluminum wire bonds. To enable silver sintering instead a gold or silver surface is recommended. In order to change the aluminum metallization on

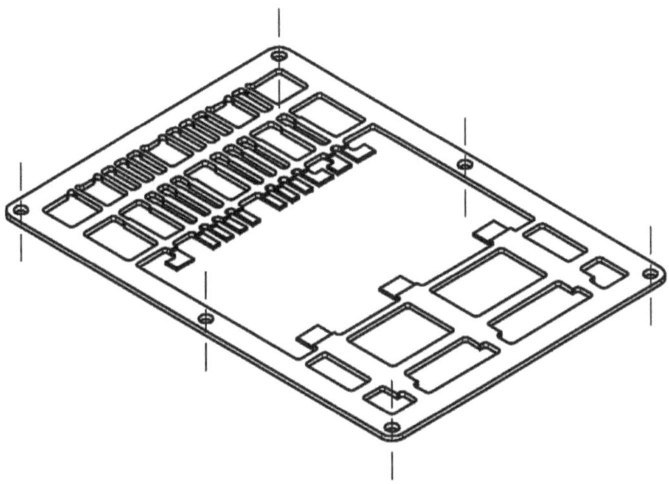

Fig. 5 The lead frame design showing the thinner parts attached between the DBC substrates

the top side of the devices while at the same time keep an adequate stand-off distance to the edge passivation of the device chips, initially an electroless plating process was considered to deposit a thick (20–30 μm) nickel and a thin (50–100 nm) immersion layer of gold on the device pads. This plating process failed and another method based on machining of the DBC copper layer was used instead [9] as explained in more detail in the Fabrication section.

3 Simulation Results

3.1 Thermal Models

Thermal models of power modules with single and double sided cooling were set up. The single sided model included SiC switching devices on a DBC substrate covered in epoxy. Results of the thermal simulation of this model are shown in Fig. 6 for a dissipated power of 65 W at two of the transistors for 180 s. The model predicted a maximum temperature of 141 °C at the center of the active devices.

The double sided model contained SiC switching devices sandwiched between two DBC substrates and electrical connections attached with sintered silver material [10]. The active SiC device chips were subjected to three power pulses with a cycle time of 3 s and an on-time of 1.5 s. The simulated power dissipation was 70 W per chip, set up as an internal heat source in two of the devices. The convection boundary conditions was set to 3000 W/(m^2 K), simulating water based heat sinks attached to the top and bottom of the power module. The results of the simulations are shown in Fig. 7. The simulated maximum temperature of the devices predicted by the model was 119.7 °C, which is 85 % of the maximum simulated temperature for the single sided version.

In addition to the large, full-scale model of the double sided cooling version, a local model containing a slice through all layers over one device chip including 0.5 mm of its surrounding materials in x and y direction was created. The local

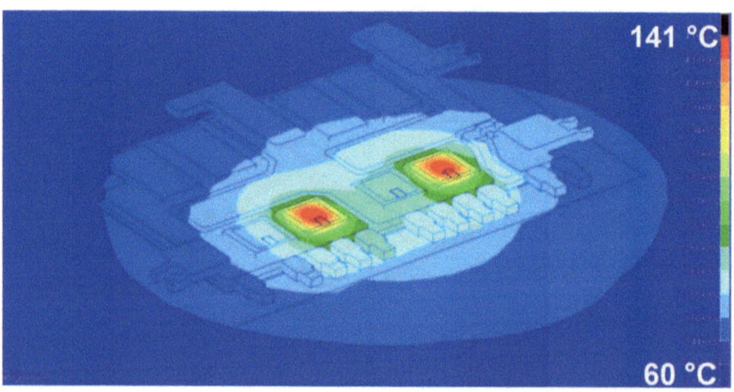

Fig. 6 Simulation results on SiC power module for active cycling after t_{on} = 180 s

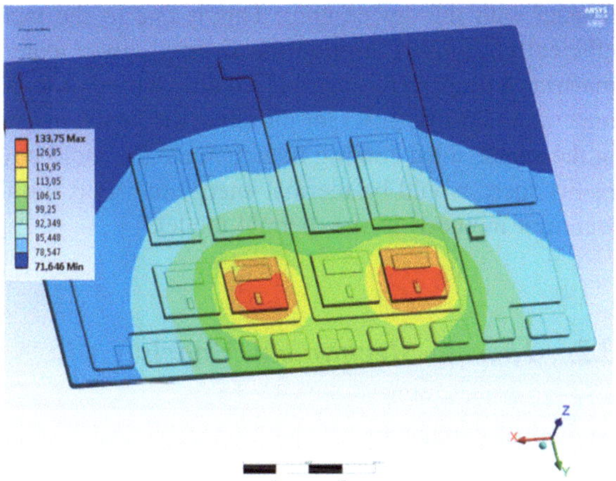

Fig. 7 Thermal simulation results of the double sided cooling version after 3 power cycles with a cycle time of 3 s and an on-time of 1.5 s

sub-model used thermal boundary conditions from the larger model on each side of the stack. Identical convection conditions on top and bottom surfaces and identical power pulses and cycle times were used. The results are shown in Fig. 8. Maximum temperature attained on the SiC device chip was 91.2 °C.

As expected, thermal simulations predict lower temperatures at the active switching devices in the models with cooling on both top and bottom. The predicted temperature decrease with double sided cooling was 15–35 %. This is not surprising since the temperature decrease is expected to be lower than 50 % because of the difference in the layout of the DBC copper layers connecting top and bottom of the device chip. This asymmetry causes a higher thermal resistance on the top side than on the bottom side of the active devices. The results of the thermal simulations in Figs. 7 and 8 show that the geometry of the epoxy molding material also influences the temperature profile. The temperature is higher where the molding material is in

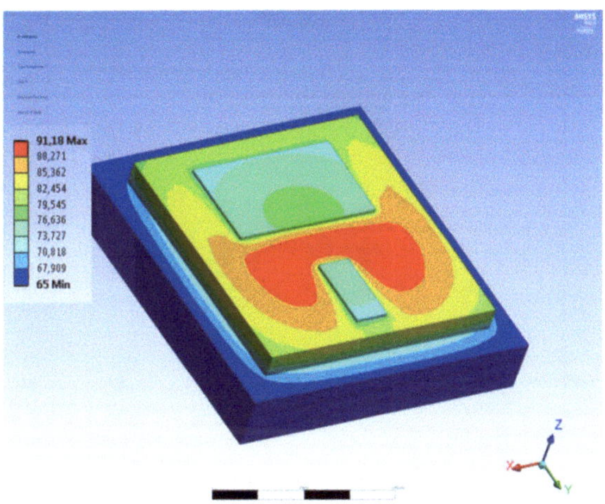

Fig. 8 Thermal simulations of double sided local model

contact with the device chip. This is due to the fact that thermal energy is more efficiently removed through the sintered silver material than through the molding material (compare the thermal conductivities of 1 W/(mK) of the epoxy molding material to 250 W/(mK) for the silver die attach material). The effect is not apparent in the single sided cooling model where the mold covers all of the device chips. In reality there would be thermal energy transport through the bond wires, but these were not included in the thermal model.

3.2 *Thermo-Mechanical Models*

Numerical simulations of thermo-mechanical response of the single sided power module were performed. The simulations were run on models of the module and a local sub-model of a bond wire interconnect shown in Fig. 9.

A typical fabrication process for the single sided module includes soldering the devices on the DBC substrate followed by transfer molding. Thermo-mechanical simulations of the assembly processes followed by a 180 s power-on and then

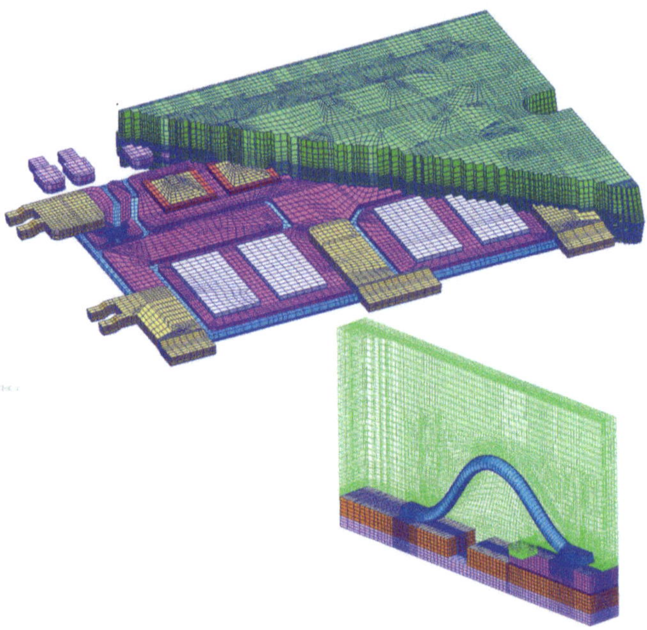

Fig. 9 Model of the single sided power module and a local sub-model of a bond wire interconnect

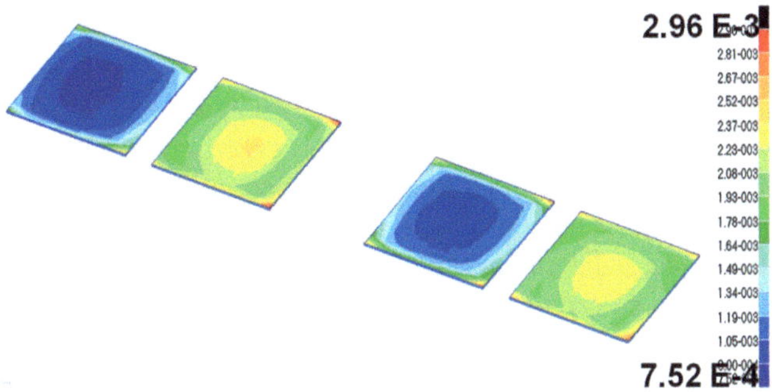

Fig. 10 Equivalent creep strain increment accumulated during one power cycle with t_{on} of 180 s

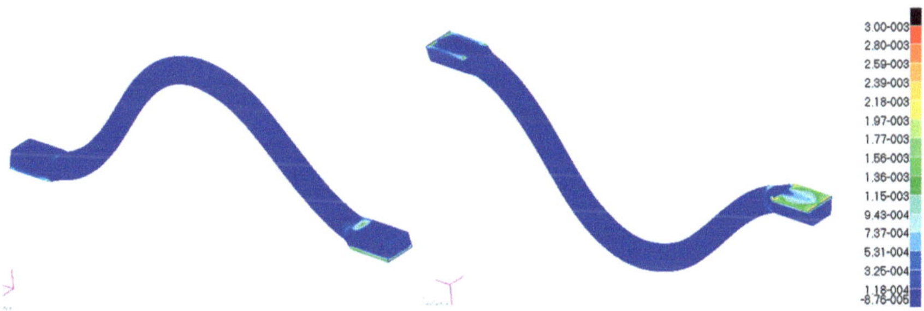

Fig. 11 Equivalent plastic strain increment accumulated in the bond wire during one power cycle with t_{on} of 180 s

cooling to room temperature were run to estimate the mechanical stresses and strains in the module during typical thermal loading. Figures 10 and 11 shows the incremental increase in creep deformation in the solder die attach and the wire bonds of the active devices after 180 s power-on and cooling down to RT. The simulated power dissipation was 65 W per device chip and the creep model used in this work for steady state creep is based on an exponential function of stress described in [11].

The thermo-mechanical creep modeling on the single sided cooling power module predicted a roughly 3 ‰ maximum rise in creep strains on both the active devices and the local wire bond after 180 s power-on, then cool down to room temperature. In general, creep strain rates during real operations will likely be larger with all switching devices powered alternately. The high CTE mismatch between DBC and SiC devices is the primary cause of significant strains and stresses evolving in the die attach. Accumulating die attach creep will lead to solder fatigue and thereby an increase of the thermal resistance to the substrate. For the local bond wire model, two major failure causes can be expected: heel cracking and bond lift, respectively. The simulation results shown in Fig. 11 indicate that the wire bond interface to the transistor metallization is the area which is most prone to generate bond lift failure due to plastic strains developing from alternating power states. A Coffin-Manson expression for the number of cycles to failure has been widely applied [12]:

$$N = a\varepsilon_{pl}^{\ -b} \tag{1}$$

It correlates the plastic strain in the damaged region with the cycles to failure by a power law with constants obtained from stress experiments. The coefficients are always related to dedicated cycling experiments and not transferrable [13].

Thermo-mechanical simulations have also been performed on the version with double sided cooling. The strain behavior is less known for sintered nanosilver particles than for the eutectic solder used in the single sided version, but has been investigated by [14–17]. Since the melting temperature of the silver is much higher than for the solder, the operational temperature of many applications is comparatively low. For example, the homologous temperature of eutectic tin-based solders at normal operating temperatures (−40 to 125 °C) ranges from typically 0.4–0.8 Tm, while for the sintered silver it ranges from 0.18 to 0.32 Tm. This suggests that creep effects should be lower in sintered silver joints compared to solder joints.

A plastic strain model from [17] and a time hardening model derived from [15] have been used to model plastic strain and primary creep strain in the sintered silver layers. The thermal loading was identical to that described in Sect. 2 with three power dissipation pulses of 70 W with a cycle time of 3 s and an on-time of 1.5 s. The stress-free temperature in the model was set to 180 °C. The results of the thermo-mechanical analysis are shown in Figs. 12 and 13.

The model predicts an equivalent (von-Mises) stress of around 160 MPa at the edges of the sintered silver layers on top of the device chip. The model predicts a small area of maximum 2.3 % strain at the corners of the sintered layers. This very

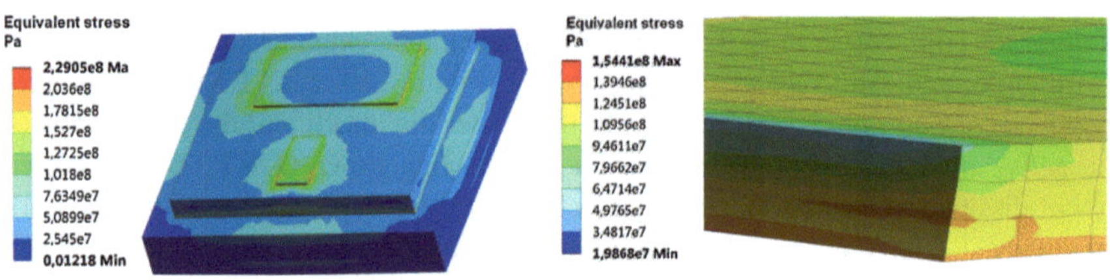

Fig. 12 Equivalent (von-Mises) stress in the sintered silver layers and a close-up of the sintered layer at the BJT base contact

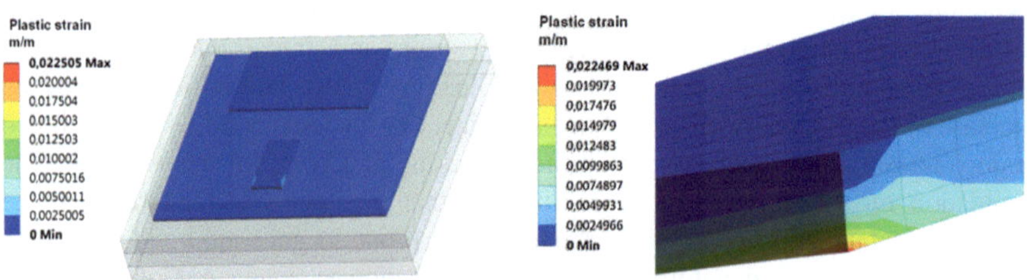

Fig. 13 Plastic strain in the sintered silver layers and a close-up of the sintered layer at the BJT base contact

high plastic strain at the corners of the sintered layers is considered to be edge effects in the simulation and not present in the real assembly. Another conclusion may be that it is beneficial to have increased porosity at the corners of the sintered layers to reduce strain in these areas. Simulated plastic strains on the top base and emitter sintered layers away from the base corners are around to 0.7 % which is a more reasonable result. The model does not predict any creep strains in the sintered layers, which is most likely a result of the short total simulated cycling time. The total cycle time was 8.85 s, including three power pulses with on-times of 1.5 s each.

4 Fabrication

The fabrication flow of the double sided power modules is shown in Fig. 14. For the nanosilver sintering either a silver or gold surfaces on the DBC and device chip pads are recommended. Since the top side interconnection is typically aluminum wirebonding, the top pads of both the BJTs and diodes are plated with aluminum. The initial plan was to use an electroless plating process to add an additional 20 μm thick layer of nickel followed by a thin (50–100 nm) layer of gold on top of the aluminum layer. This would prepare the surface for the nanosilver sintering process and, in addition, allow for enough stand-off distance to the edge passivation.

After the plating attempts, the device chips were molded in epoxy and micro-sectioned for SEM analysis. Figure 15 shows SEM images of the cross-sectioned results for two of the plating attempts. The first one was with the original plating process recipe. It can be seen that delamination between the aluminum and the nickel layers occurs. The second SEM image is from a sample with a modified recipe which allows shorter time for nickel growth. The result is a thinner nickel layer but better adhesion to the aluminum layer.

The conclusion of the plating experiments is that the plating process could not be used to form the stand-off distance to the edge passivation. It does, however, provide an Au surface for the nanosilver sintering process.

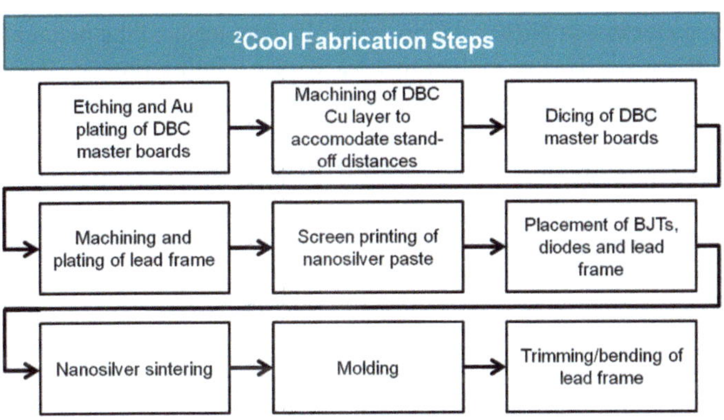

Fig. 14 Fabrication flow of the double-sided cooling version (named ²Cool)

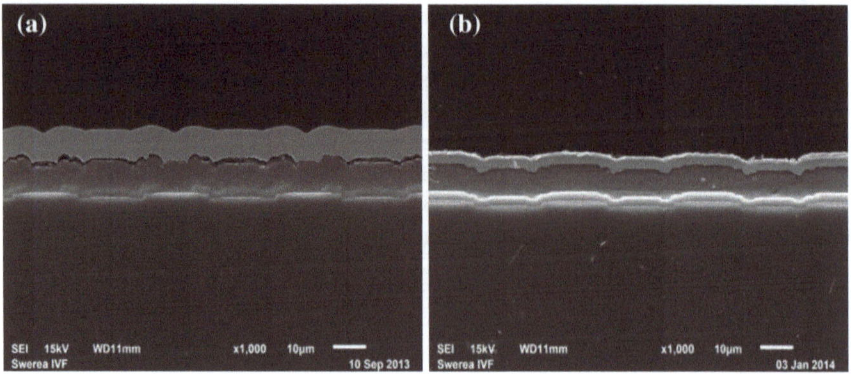

Fig. 15 **a** Results with original plating process recipe and **b** with modified plating process

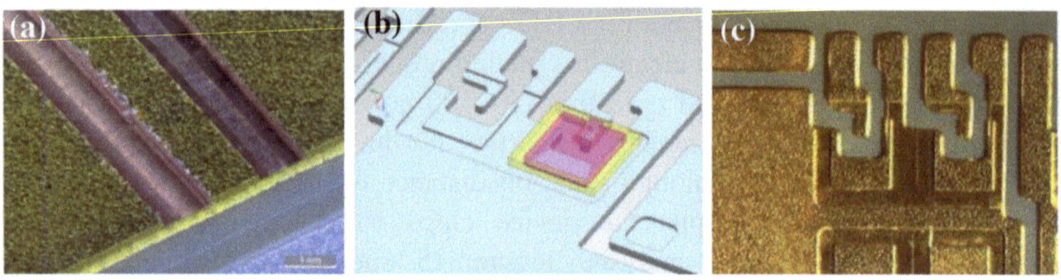

Fig. 16 **a** Machined groove as a feasibility test (with machining fluid on the groove to the *right*), **b** the design of the machined areas of the *top-side* DBC substrate, and **c** the machined DBC substrate

An alternative method to achieve the stand-off distance is to machine a groove in the copper layer in the area of the die edges. This creates a plateau of copper on which the die is sintered with the nanosilver. Figure 16 shows a feasibility test of a 0.1 mm deep groove machined into a 0.3 mm thick copper layer of a DBC. This is what will be machined in the top DBC around the BJT and diode die locations.

Following electroless Ni/Au plating of the die contacts and the preparation of the top side DBC, nanosilver is applied using a screen-printing process. The die and a specially designed lead frame are placed on the nanosilver and sandwiched by the bottom DBC substrate.

A sintering process in a furnace at 250 °C with pressure applied at the peak temperature is then performed. The peak temperature and pressure is maintained for 5 min. The resultant joining interface is then expected to be operational up to at least 0.6 times the melting temperature of Ag at Tm = 961 °C. Adding pressure during the sintering process is an important step, which improves the quality of the sintered joint.

After the sintering process the module is filled with a non-conductive epoxy and vacuum molded. The fabricated module is shown in Fig. 17.

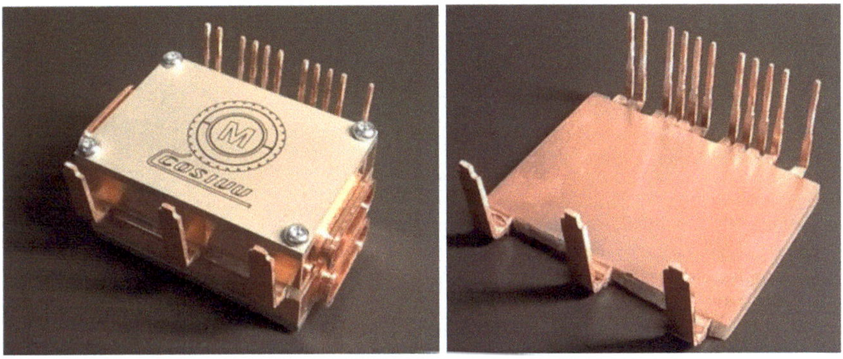

Fig. 17 The double sided ^{2}Cool module with and without attached heat exchangers

5 Summary

The design and fabrication process of a power module with double sided cooling has been presented. Each double sided module consists of four transistors, and four diodes. The device chips are sandwiched between two DBC substrates and the electrical connection on the top and bottom side are made with sintered nanosilver. Nickel/gold plating of the top side contacts is required for the nanosilver sintering process. A new method to create a stand-off distance to the edge passivation of the chips has been used.

Thermal simulation models predicted a reduction of maximum temperature on the active switching devices from 141 °C to between 91.6 and 119.7 °C when utilizing double sided cooling. Creep modeling on the single sided cooling power module showed that roughly 3 ‰ maximum rise in creep strains on both the active devices and the local wire bond can be expected. The double sided cooling thermo-mechanical model predicted plastic strains at the sintered silver layers of 0.7 % and no creep strains. This is believed to be a result of the short thermal loading in the model and is expected to change in an expanded model, which adequately simulate the individual processing steps and a more severe thermal loading.

In conclusion, simulations predict that double sided cooling of power modules can lower the operational temperature by between 15 and 35 %, which will lead to smaller or fewer devices for a given power requirement specification. Or, alternatively improve reliability margins and increase life-time of the power electronics devices and packaging.

The targets in power electronics, particularly within automotive electric drive applications, are higher switching frequencies and current densities in combination with smaller system volumes and lower cost. Effectively removing dissipated heat from the switching devices enables a higher current carrying capability per chip area ratio, thus leading to smaller or fewer devices for a given power requirement specification. One way to achieve a larger heat exchange area and thereby a reduction in thermal resistance is to cool both sides of the devices.

Acknowledgments The Authors would like to acknowledge the European Commission for supporting these activities within the COSIVU project under grant agreement number 313980.

References

1. Gustafsson T, Nord S, Andersson D, Brinkfeldt K, Hilpert F (2014) COSIVU—compact, smart and reliable drive unit for commercial electric vehicles. In: Proceedings of AMAA 2014, advanced microsystems for automotive applications 2014, lecture notes in mobility. Springer, pp 191–200
2. Otto A, Kaulfersch E, Brinkfeldt K, Neumaier K, Zschieschang O, Andersson D, Rzepka S (2014) Reliability of new SiC BJT power modules for fully electric vehicles. In: Proceedings of AMAA 2014, advanced microsystems for automotive applications 2014, lecture notes in mobility. Springer, pp 235–244
3. Weitzel CE, Palmour JW, Carter CH Jr, Moore K, Nordquist KJ et al (1996) Silicon carbide high-power devices. IEEE Trans Electron Dev 43(10):1732–1741
4. Cooper JA Jr, Agarwal A (2002) SiC power-switching devices—the second electronics revolution? Proc IEEE 90(6):956–968
5. Chang H-R, Bu J, Kong G, Labayen R (2011) 300A 650 V 70 um thin IGBTs with double-sided cooling. In: Proceedings of the international symposium on power semiconductor devices and ICs, pp 320–323
6. Ning P, Liang Z, Wang F (2013) Double-sided cooling design for novel planar module. In: Conference proceedings—IEEE applied power electronics conference and exposition—APEC, pp 616–621
7. Zhang H, Ang SS, Mantooth HA, Krishnamurthy S (2013) A high temperature, double-sided cooling SiC power electronics module. In: 2013 IEEE energy conversion congress and exposition, ECCE 2013, pp 2877–2883
8. Hoene E, Ostmann A, Lai BT, Marczok C, Müsing A, Kolar JW (2013) Ultra-low-inductance power module for fast switching semiconductors. In: PCIM Europe conference proceedings, pp 198–205
9. Brinkfeldt K, Edwards M, Andersson D, Neumaier K, Zschieschang O et al (2015) Modeling and fabrication of a SiC-based power module with double sided cooling. In: 20th annual pan pacific microelectronics symposium 2015
10. Bai JG, Zhang ZZ, Calata JN, Lu G-Q (2006) Low-temperature sintered nanoscale silver as a novel semiconductor device-metallized substrate interconnect material. IEEE Tran Compon Packag Technol 29(3):589–593
11. Schubert A, Walter H, Dudek R, Michel B, Lefranc G et al (2001) Thermo-mechanical properties and creep deformation of lead-containing and lead-free solders. In: Proceedings of the international symposium and exhibition on advanced packaging materials processes, properties and interfaces, pp 129–134
12. Chidambaram NV (1991) A numerical and experimental study of temperature cycle wire bond failure. Paper presented at the proceedings—electronic components conference, pp 877–882
13. Ramminger S, Seliger N, Wachutka G (2000) Reliability model for al wire bonds subjected to heel crack failures. Microelectron Reliab 40(8–10):1521–1525
14. Chen G, Sun X-U, Nie P, Mei Y-H, Lu G-Q, Chen X (2012) High-temperature creep behavior of low-temperature-sintered nano-silver paste films. J Electron Mater 41(4):782–790
15. Chen G, Zhang Z-S, Mei Y-H, Li X, Lu G-Q, Chen X (2013) Ratcheting behavior of sandwiched assembly joined by sintered nanosilver for power electronics packaging. Microelectron Reliab 53(4):645–651

16. Li X, Chen G, Wang L, Mei Y-H, Chen X, Lu G-Q (2013) Creep properties of low-temperature sintered nano-silver lap shear joints. Mater Sci Eng A, 579:108–113
17. Dudek R, Döring R, Sommer P, Seiler B, Kreyssig K et al (2014) Combined experimental-and FE-studies on sinter-Ag behavior and effects on IGBT-module reliability. In: 15th international conference on thermal, mechanical and multi-physics simulation and experiments in microelectronics and micro-systems, EuroSimE 2014

Performance Evaluation of Permanent Magnet Assisted Synchronous Reluctance Motor for Micro Electric Vehicle

Bogdan Varaticeanu, Paul Minciunescu and Silviu Matei

Abstract A 7.5 kW Permanent Magnet Assisted Synchronous Reluctance Motor (PMASynRM) that works with ferrite magnets has been designed for a micro electric vehicle. Finite element analysis is used to evaluate machine performances and to study the demagnetization effect of ferrite PMs. An experimental validation of the PMASynRM characteristics is made. The measured values of the back electromotive force and torque are compared with the numerical predicted values. The PMASynRM electromagnetic losses are investigated in order to increase the motor efficiency.

Keywords Permanent magnet assisted synchronous reluctance motor · Ferrite magnet · Demagnetization · Efficiency · Reluctance · Finite element method · Traction motor

1 Introduction

Permanent magnet (PM) assisted synchronous reluctance motors are an interesting choice for traction application, due to their high torque, high efficiency and capability to operate in a wide speed range. This machine generates torque from saliency

B. Varaticeanu (✉)
Servomotors Departament and Politehnica University of Bucharest, ICPE,
Splaiul Unirii 313, 030138 Bucharest, Romania
e-mail: bogdan.varaticeanu@icpe.ro

P. Minciunescu · S. Matei
Servomotors Departament, ICPE, Splaiul Unirii 313, 030138 Bucharest, Romania
e-mail: pmagnetics@gmail.com

S. Matei
e-mail: silviu.matei@icpe.ro

T. Schulze et al. (eds.), *Advanced Microsystems for Automotive Applications 2015*,
Lecture Notes in Mobility, DOI 10.1007/978-3-319-20855-8_14

and by adding PMs in the rotor flux barriers an additional torque component is produced. Being largely dependent on reluctance torque, PMASynRM is suitable to replace the traditionally expensive rare earth PMs with ferrite PMs without dramatically affect efficiency, power factor and torque density [1, 2]. The equation of torque production in PMASynRM is given by [3]:

$$T = T_r + T_m = \frac{3}{2}p(L_d - L_q)I_d I_q + \frac{3}{2}p\Psi_m I_q \tag{1}$$

where T_r, T_m are the reluctance torque component and magnet torque component, p is the number of pole pairs, Ψ_m is the magnitude of flux of the magnets, L_d, L_q, I_d and I_q are the d and q axis inductances and currents respectively.

The design and optimization of the proposed 7.5 kW PMASynRM is described in [4]. The rotor structure of the machine is analyzed regarding the number of poles and shape of the flux barrier in order to obtain high torque and high power. The stator structure is examined regarding the cogging torque simulated at no-load condition and torque ripple simulated at nominal load condition. The proposed PMASynRM configuration is based on a 36-slots stator and a 6 poles rotor. In Fig. 1 the rotor and stator parts of the 7.5 kW proposed machine is shown.

The rotor geometry satisfies the electromagnetic design criteria together with mechanical design criteria regarding the mechanical stress in the rotor parts. The stator geometry and winding configuration satisfy the electromagnetic design criteria and meet the cost reduction in the manufacturing process by using a conventional induction motor lamination. In Table 1 the specifications of the proposed 7.5 kW PMASynRM is shown.

This paper is dealing with numerical and experimental performance evaluation of the final machine configuration.

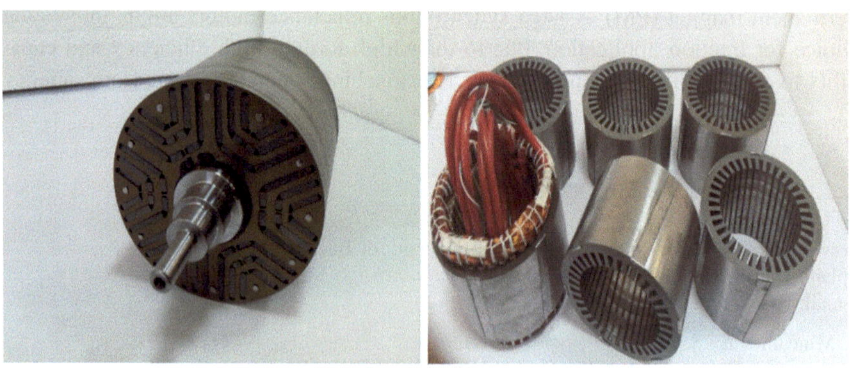

Fig. 1 Rotor and stator parts of the 7.5 kW PMASynRM: rotor (*left*); stator (*right*)

Table 1 PMASynRM motor specification

Dimension/rating	Value
Stator slots	36
Number of magnetic poles	6
Stator outer diameter	170 mm
Rotor outer diameter	114 mm
Airgap length	0.5 mm
Stack length	110 mm
Rated voltage	100 V
Rated current amplitude	101.4 A
Maximum speed	9000 rpm
Threshold speed	2100 rpm
Remanent flux density of ferrite PM	0.4 T
Number of phase	3
Winding type	Distributed

2 Performance Evaluation

The design characterization of the proposed PAMSynRM is made, both by using Finite Element Analysis and experimentally. The most important machine parameters like back electromotive force and electromagnetic torque experimentally obtained are compared with the numerically predicted values in order to evaluate the numerical model results.

2.1 Numerical Evaluation of the PAMSynRM Performance

For an accurate prediction and for optimization of PAMSynRM performance, finite element method (FEM) is used. The structure of the proposed machine was optimally design and is evaluated base on the 2-D and 3-D FEM. The nonlinear magnetic materials chosen in the machine design and complexity of the machine geometry impose for skewed stator 3D numerical simulation in order to investigate the machine performance. In order to reduce the cogging torque and the torque ripple, the stator slots were skewed with one slot. In Fig. 2 the magnetic flux density distribution in the proposed machine, simulated by 2D and 3D FEM is shown.

In Fig. 3a the back electromotive force (EMF) of one phase at rotor speed of 1000 rpm, simulated using: 2D FEM for unskewed stator and 3D FEM for skewed stator is shown. It is observed that the waveform of the electromotive force change to an almost sinusoidal shape in the skew stator. The average value of the 3D predicted back EMF has a slight decrease of 6.2 % compare with 2D predicted value, as a consequence of skewing stator. In Fig. 3b the cogging torque waveform, numerically predicted for the stator with skewed and unskewed slots is shown. Cogging torque of PMASynRM is the torque due to the interaction between ferrite

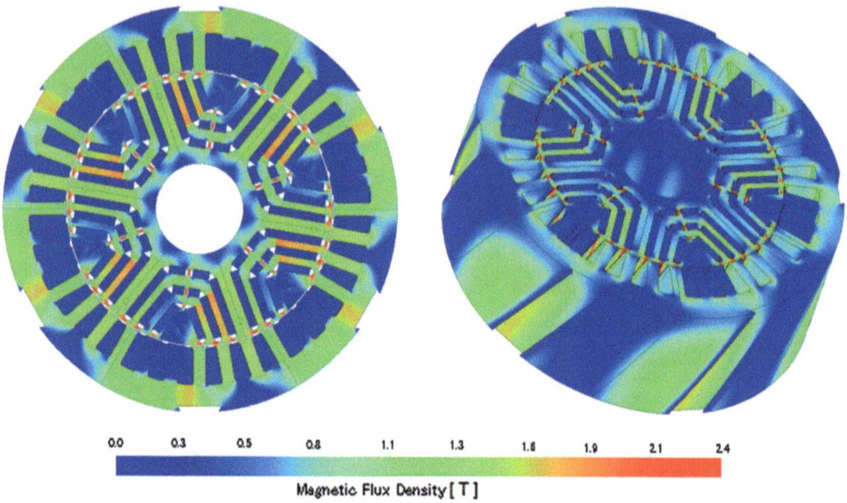

Fig. 2 Magnetic flux density distribution in PAMSynRM: 2D and 3D simulated

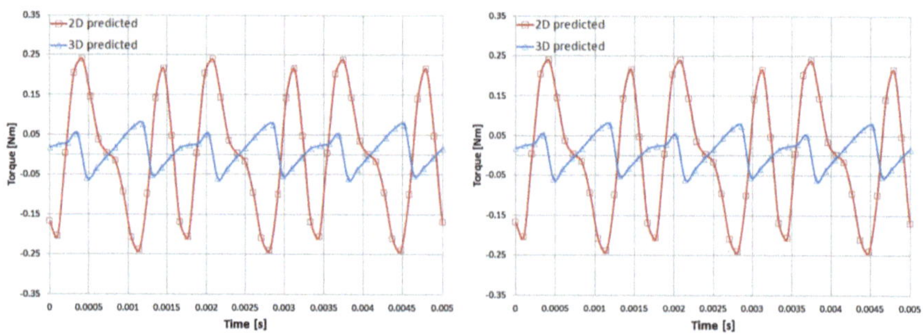

Fig. 3 Machine performance prediction at no-load: back EMF (*left*); cogging torque (*right*)

PMs on the rotor and the stator teeth, when the motor is at no-load condition. It is observed that the skewed stator has a major reduction in cogging torque (83 %), compared with unskewed stator.

A common drawback of the PAMSynRM is the high torque ripple. The optimization of the rotor and stator structures were done in order to obtain high torque and high power in accordance with low cogging torque and low torque ripple. Many approaches to reduce the torque ripple were studied in the design and optimization process [4], and was concluded that one of the most efficient method is skewing the stator slots. In this type of machine, the maximum torque will occur when the current is shifted by an angle (current phase angle) relatively to the q axis. When the motor operates in the constant torque region, the current phase angle for

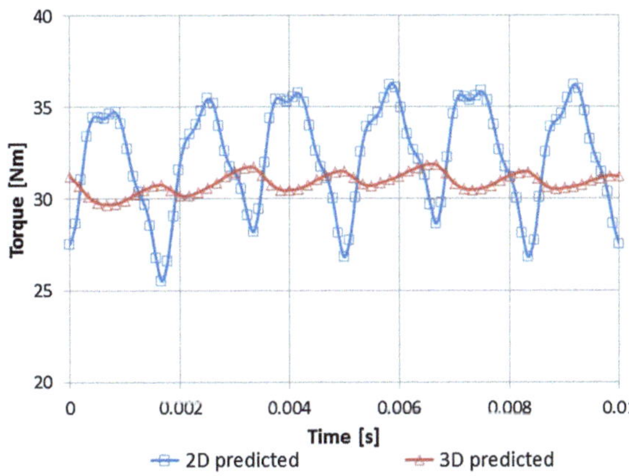

Fig. 4 2D and 3D predicted electromagnetic torque waveform

which the average torque has the maximum value is 45°. In Fig. 4 the electromagnetic torque waveform, simulated using: 2D FEM for unskewed stator and 3D FEM for skewed stator is shown. It is observed that the skew stator has a major reduction in torque ripple (5 %), compared with unskew one (28 %), This high advantage come with a small inconvenience, the value of the average torque for skewed stator has a slight decrease of 4 % compared with unskewed stator.

In PAMSynRM torque production is characterized by two components: the reluctance torque due to saliency and the PM torque due to the interaction between PMs and stator currents. In Fig. 5 the torque and output power versus speed curves is shown. Using the equation of torque production (1) the torque waveform is decomposed in reluctance component and PM's flux component. It is observed that

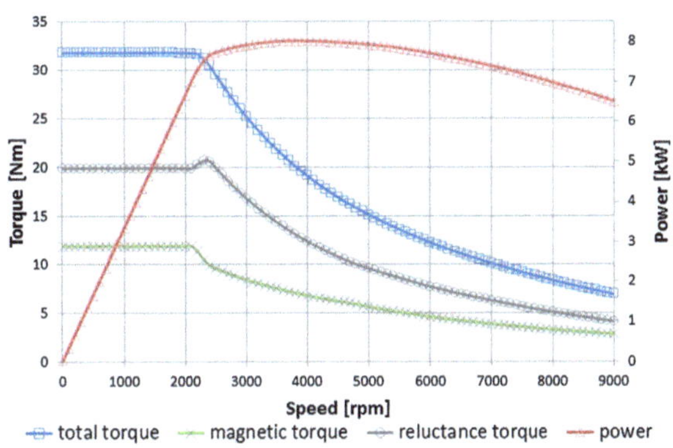

Fig. 5 Torque and output power versus speed

this type of machine has two distinct regions of operation: the constant torque operation below the threshold speed and the constant power above the threshold speed. The threshold speed is defined as the maximum speed up to which the rated torque can be maintained constant.

The predicted rated torque in the constant torque region has a value of 31.5 Nm and after threshold speed (2100 rpm) this gradually decreases up to 7 Nm. This motor has an extended operation range due to wide field weakening region, reaching a maximum speed of 9000 rpm. In the constant torque region, the maximum torque value will be maintained until the maximum power is reached. The maximum output power of this motor is 7.5 kW and is kept constant between 2100 and 6500 rpm. After 6500 rpm the electromagnetic losses induced in the stator and rotor cores have a substantial increase leading to a drop in power. The waveform of the total torque versus speed was numerically computed using FEM, however the values of magnetic torque and reluctance torque are only estimated. In the constant torque region, the reluctance torque seems to be over 65 % of the total torque.

In Fig. 6 the variation of induction characteristics due to the current phase angle is shown, taking in consideration the rotor and stator cores saturation.

The inductance simulation is done at the threshold speed when through the coil winding flow the rated current amplitude. The maximum saliency ratio is 5 and is obtained for a phase angle of 75°. In this type of machine, choosing an appropriate current phase angle it is possible to increase the torque and power production and to improve the efficiency. For the phase angle of 45° (when average torque has the maximum value) the saliency ratio is 4.5.

The electromagnetic losses that affect the PMASynRM performance and efficiency are: the resistive losses (cooper losses) experienced by the winding when the current flows through the coils, and the iron losses experienced by stator and rotor cores due to time variation at the magnetic flux that links the magnetic circuit. In this machine, due to field weakening control, it is possible to speed up the rotation, leading to an increase frequency of fluctuation of the magnetic field applied to the

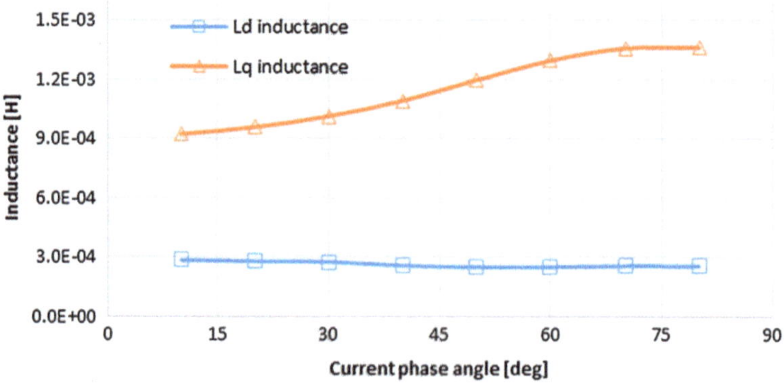

Fig. 6 L_d and L_q inductances versus current phase angle

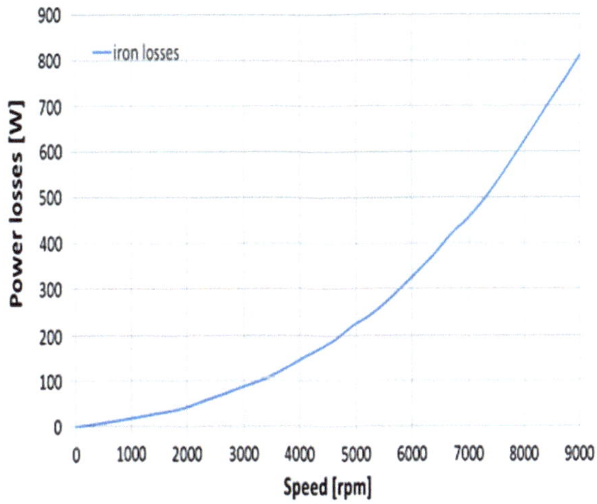

Fig. 7 Iron losses of the PMASynRM

stator and rotor cores, increasing also the iron losses induced in the machine magnetic circuit. The ferrite PM due to the low electrical conductivity does not experience any magnet losses.

In Fig. 7 the iron losses function of speed, simulated in various points defined by the torque vs. speed curve from Fig. 5 is shown. The losses analysis was made taking into consideration the worst scenario that can occur in the machine operating regime: the rated current amplitude flow through the coil winding having a temperature of 145 °C, and the speed is variable. In order to increase the motor efficiency, in the design and optimization process is chosen for stator core a lamination of 0.35 mm thickness. From technical reasons for rotor core a lamination of 0.5 mm thickness is chosen. It is reinforce that iron losses increase exponentially with speed reaching a maximum value of 800 W at 9000 rpm. In Fig. 8 the efficiency map of the proposed PMASynRM is shown.

The efficiency map is simulated at rated current amplitude of 102 A and rated voltage of 100 V. The machine exhibits a high efficiency over a wide operation range (90 % efficiency over 60 % of operating range and 95 % efficiency over 18 % of operating range), reaching a maximum efficiency of 96 %.

2.2 *Experimental Evaluation of the PMASynRM Performance*

The experimental validation of the proposed machine performance is made using the test bench shown in Fig. 9.

The machine mounted on a test bench is shown in Fig. 10. The line to line back EMF at 1000 rpm, measured and predicted by 3D FEM is shown in Fig. 11. Even if the predicted voltage waveform of the EMF was obtained by means of 3D FEM,

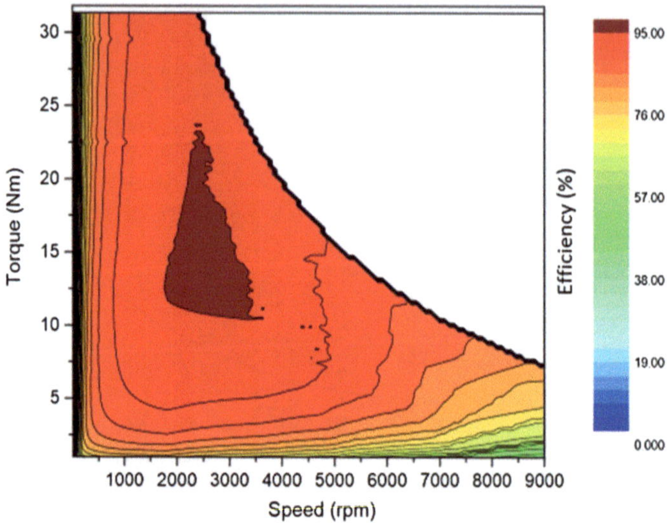

Fig. 8 Efficiency of the PMASynRM

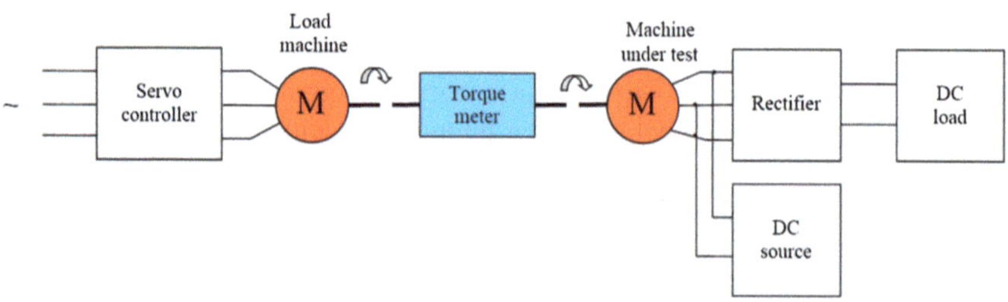

Fig. 9 Test bench used for experimental validation of the PMASynRM performance

Fig. 10 PMASynRM mounted on a test bench

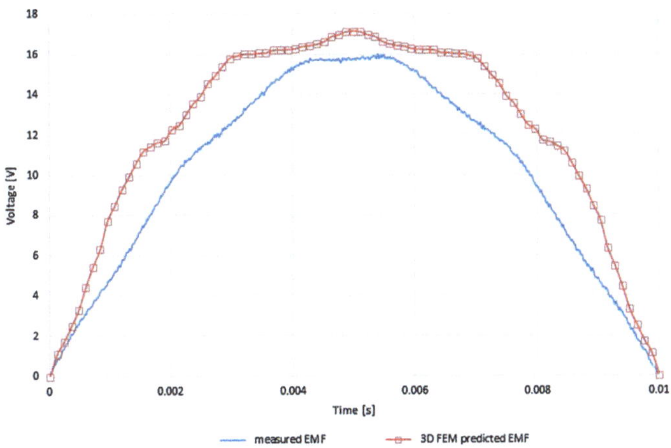

Fig. 11 Line to line experimental back EMF and 3D predicted by FEM

there is a slight increase in amplitude compared to measured waveform. The values of the predicted voltage are higher than the measured values.

The cogging torque also known as "no-current" torque is measured with a bar mounted on the machine under test shaft and a force gauge meter. The pick value of 0.4 Nm was measured, which is less than 2 % from nominal torque value. This experimental measured value is characterized by two components: the cogging torque component and the friction torque in the bearings component. The FEM predicted value of the cogging torque do not consider the bearings friction torque, as a result the 3D FEM predicted value and experimental measured value are different.

The electromagnetic torque of the PMASynRM machine is measured using a static method. The terminals of the machine under test are supplied with a current from a DC power supply. The current projections to the vertical coordinate represent their instantaneous current values [5], which are shown in Fig. 12a. A DC power supply is used to feed the needed currents for the machine three phases,

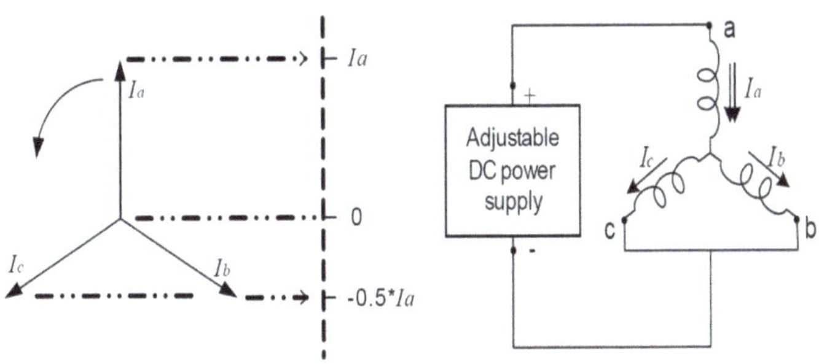

Fig. 12 Machine phase currents at static torque measurement: instantaneous current values (*left*); circuit diagram in static torque measurement (*right*)

Fig. 12b. At the same current amplitude, the torque value can be adjusted by changing the current phase angle. The current phase angle for the currents shown in Fig. 10a is zero. Creating an imbalance in the phase currents, a non-zero current phase angle can be simulated. By inserting a variable resistance in series with phase B, the current phase angle is modified.

For A phase current of I_a = 101.4 A, B phase current of I_b = −23.9 A and C phase current of I_c = −77.5 A, the current phase angle is 45°. At this value of the currents phase angle and at the rated current amplitude, the measured electromagnetic torque has a value of 31 Nm. This value is approximatively equal with the 3D predicted value, confirming in this way the design value.

3 Demagnetization Analysis

In PM motors the temperature and the magnetic field inside de magnets need to be kept under control in order to avoid irreversible demagnetization. Ferrite magnets experience demagnetization phenomenon due to low temperature and external magnetic fields that push the operating point of the magnet below the knee point. In Fig. 13 the demagnetization curves at 20 and −40 °C of ferrite PM's type Y30H used in the proposed PMASynRM is shown. The knee point at 20 °C appears in third quadrant and when the temperature decreases at −40 °C the knee point moves in the second quadrant. The critical temperature considered in this analysis, that imposes the worst condition in irreversible demagnetization is −40 °C, which is the lowest temperature assumed in the usual automotive application. At temperature of −40 °C when the operating point go below the knee point the ferrite magnet

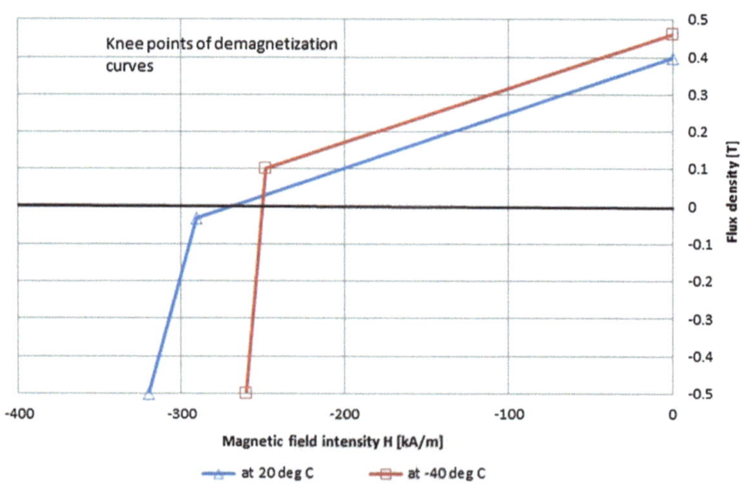

Fig. 13 Demagnetizations curves of ferrite PM at 20 and −40 °C

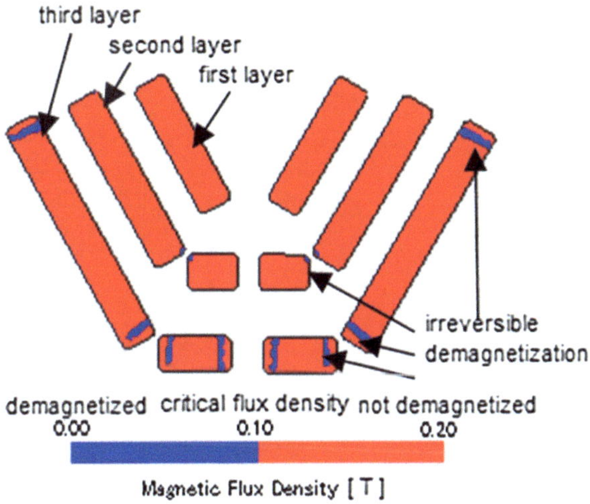

Fig. 14 Magnetic flux density in ferrite PM at −40 °C and current amplitude 280 A

suffer an irreversible demagnetization. The critical flux density below which the demagnetization becomes irreversible is assumed to be 0.1 T, as shown in Fig. 13.

In order to evaluate the demagnetization, the magnetic flux density in the ferrite PM is calculated. To assume the worst condition in operation of PMASynRM, the temperature is set to −40 °C and the magnetic flux density due to the stator winding is set to flow in opposition with magnets flux [6]. In Fig. 14 the magnetic flux density in one magnetic pole of the ferrite PM is shown, when through the coil winding flow a current amplitude of 280 A (three times higher than the amplitude of rated current). The contour plot is shown only in two colours in order to emphasis the critical value of the flux density (0.1 T), below which in ferrite magnets will take place an irreversible demagnetization. As it can be observed, demagnetization occurs in a lot of parts of the magnets in the second and third layers. The first layer of magnets is the layer near the rotor surface, the third layer of magnets is the layer near the rotor shaft and the second layer of magnets is the layer placed between the first and second layers.

The demagnetization rate experienced by each layer of magnets is shown in Fig. 15. This parameter is the rate of irreversible demagnetization calculated as magnetic flux density influenced by external factors (temperature and opposing magnetic field) against the initial magnetic flux density. The demagnetization rate is examined up to a current amplitude of 3 times higher than rated current amplitude. It is observed that starting with the current amplitude of 220 A the irreversible demagnetization occurs in the third layer of ferrite PM, and the demagnetization ratio increase exponentially with the current. The first and second layers of ferrite PM are not sever affected by demagnetization even if the current amplitude is 300 A. The rate of demagnetization due to an opposing magnetic field is influenced by the power of this field. The temperature is another parameter that affects the rate of demagnetization. If for rare earth PM the highest temperature influence the rate of demagnetization, for ferrite PM the lowest temperature is critical.

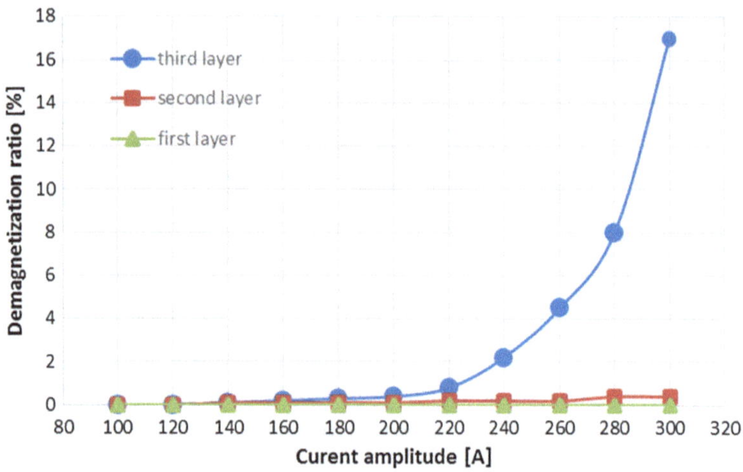

Fig. 15 Demagnetization ratio

The back electromotive force of the one phase coil at 1000 rpm over one period of electric angle, before and after demagnetization is shown in Fig. 16. At current amplitude of 200 A, the demagnetization occured in ferrite PM is reversible and does not influence in the negative way the performance of the PMASynRM. In contrast, at a current amplitude of 280 A the irreversible demagnetization cause changes of the back electromotive force waveform, decreasing the motor performance.

Due to the motor configuration (36 slots and 6 magnetic poles), the third layer of ferrite magnets are first subjected to irreversible demagnetizations when the current amplitude is increasing and motor is working in worst-case scenario. Due to the commutation process, through coil winding can flow for very short period of time,

Fig. 16 Phase back electromotive force before demagnetization and after demagnetization

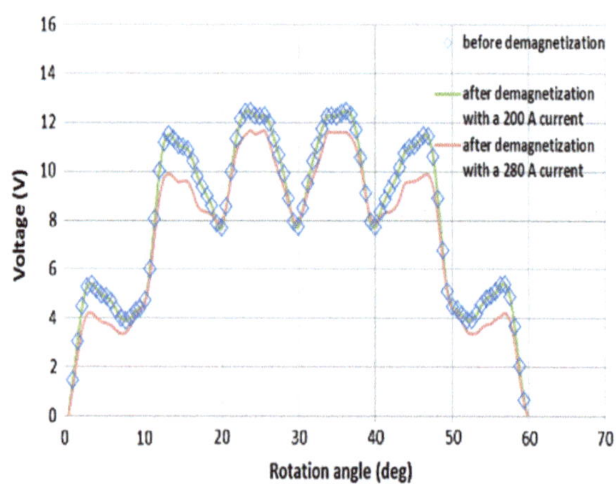

currents with higher amplitude than the rated current. Very high current amplitude (3 times higher than rated current) combined with low temperature (−40 °C) can cause irreversible demagnetization in some parts of the ferrite PMs.

4 Conclusions

This paper describes the modelling and simulation of a permanent magnet assisted synchronous reluctance motor that work with ferrite PMs that will be part of the powertrain of a Micro Electric Vehicle (EV).

Performance evaluation of a 7.5 kW PMASynRM with ferrite PMs is presented. A 2D and 3D FEM approach was used to analyse the motor characteristics. The configuration of the studied machine requires 3D finite element analyses for accurate performance prediction.

Due to their high torque, high efficiency and capability to operate in a wide speed range, the PMASynRM are attractive choices for automotive industry. By replacing rare-earth PMs with ferrite PMs, a compact, low-cost traction motor is obtained. The price paid for these advantages is a small reduction in efficiencies, power factors and torque densities compared to permanent magnet synchronous motor.

The effect of low temperature and opposing magnetic field was analysed in order to evaluate the effect of irreversible demagnetization that occurs in ferrite PM. The motor configuration imposed that the layer of magnets located near the rotor shaft is first subjected to the demagnetization. The irreversible demagnetization occurs in the ferrite PMs at a temperature of −40 °C and a current amplitude of 2.8 times higher than rated current amplitude. For automotive industry and especially for this particular case the irreversible demagnetization is unlikely to take place in ferrite PMS.

The PMASynRM achieved a high efficiency (over 90 %) across a wide operation range. The iron losses have an exponentially variation function of speed, becoming the dominant source that influence the machine efficiency at high speed.

The numerical and experimental results of the proposed machine have been compared in order to confirm the design value.

Acknowledgement This work was supported by the European Commission through the Seventh Framework Programme (FP7), GO4SEM Grant Agreement No. 609256/2013 and PlusMoby Grant Agreement No. 605502/2013 and by the Sectoral Operational Programme Human Resources Development 2007–2013 of the Ministry of European Funds through the Financial Agreement POSDRU/159/1.5/S/134398.

References

1. Montalvo-Ortiz E, Foster SN, Cintron-Rivera JG, Strangas EG (2013) Comparison between a spoke-type PMSM and a PMASynRM using ferrite magnets. In: IEEE international electric machines and drives conference (IEMDC)
2. Ooi S, Morimoto S, Sanad M, Yukinori Inoue Y (2013) Performance evaluation of a high-power-density PMASynRM with ferrite magnets. IEEE Trans Ind Appl 49(3)
3. Chen X, Wang J, Lazari P, Chen L (2013) Permanent magnet assisted synchronous reluctance machine with fractional-slot winding configurations. In: IEEE international electric machines and drives conference (IEMDC)
4. Varaticeanu BD, Minciunescu P, Matei S (2014) Design of permanent magnet assisted synchronous reluctance motor for light urban electric vehicle. In: International symposium on fundamentals of electrical engineering, 27–29 Nov 2014
5. Speed PC-BDC User's manual
6. Refaie AMEL, Alexander JP, Galioto S, Reddy PB, Huh KK, de Bock P, Shen X (2014) Advanced high-power-density interior permanent magnet motor for traction applications. IEEE Trans Ind Appl 50(5)

Comparison of Energy Optimization Methods for Automotive Ethernet Using Idealized Analytical Models

Stefan Kunze, Rainer Pöschl and Andreas Grzemba

Abstract With the number of Automotive Ethernet implementations increasing, the energy efficiency will eventually become an issue. While there have been various methods proposed, their implementation is not yet the main focus. In this paper two of these approaches are considered and analytically compared using idealized network traffic models. To accommodate different applications, various traffic models are considered and the influence of different parameters on the power consumption of an Electronic Control Unit is analyzed.

Keywords Automotive ethernet · Energy efficiency · Energy Efficient Ethernet · Low Power Sleep

1 Introduction

Automotive Ethernet is an emerging technology. A wide range of in-car applications predicted to utilize Ethernet systems, such as BroadR-Reach, in the future. These applications reach from adoption in new Advanced Driver Assistance Systems (ADAS) to the replacement of currently used communication systems like Low-Voltage Differential Signaling (LVDS) or Media Oriented Systems Transport (MOST) [1]. With the number of Automotive Ethernet nodes rising, their energy

S. Kunze (✉) · R. Pöschl
Deggendorf Institute of Technology, Technology Campus Freyung,
Grafenauer Straße and Edlmairstraße 22, 94078 Freyung, Germany
e-mail: stefan.kunze@th-deg.de

R. Pöschl
e-mail: rainer.poeschl@th-deg.de

A. Grzemba
Deggendorf Institute of Technology, Edlmairstraße 6+8, 94469 Deggendorf, Germany
e-mail: andreas.grzemba@th-deg.de

T. Schulze et al. (eds.), *Advanced Microsystems for Automotive Applications 2015*,
Lecture Notes in Mobility, DOI 10.1007/978-3-319-20855-8_15

efficiency will eventually become an issue. While various methods for the energy optimization of Automotive Ethernet have been proposed, their adoption into series production is not yet the main focus.

2 Energy Optimization Methods

Strict emission goals force the automotive industry to consider even small amounts of preserved energy as worthwhile contributions. For this reason, the energy efficiency of the bus systems responsible for the communication between the Electronic Control Units (ECUs) has become a focus [2]. Various technologies for the energy optimization of these systems have been developed. For example, Partial Networking was introduced for the Controller Area Network (CAN). This technology allows hibernating ECUs and waking them selectively via special bus messages. Similar considerations have been proposed for other bus systems, such as for FlexRay [3].

2.1 Energy Efficient Ethernet

Energy Efficient Ethernet (EEE) is a technology, which allows reducing the power consumption of idle Ethernet links. It is specified in the IEEE 802.3 standard [4]. Instead of transmitting a continuous stream of idle symbols, the Physical Layer Entities (PHYs) of inactive links may enter the so called Low Power Idle (LPI) mode. In this mode the link partners seize transmission, thus reducing their power consumption. In order to assure the integrity of the link, short refresh pulses are transmitted periodically during the LPI mode. A new linkup procedure is therefore not necessary to resume normal communication. For 100BASE-TX the duration T_q, where the link is quiet may be up to 110 time longer than the time T_r, which is required for transmitting a single refresh pulse [4]. The basic principle of the EEE operation is illustrated in Fig. 1.

The power consumption of a PHY in LPI mode is approximately 10 % of the power consumption when transmitting data or idle symbols [5]. The maximum allowed transition time T_{EEE}, required to enter and leave the LPI mode, is specified

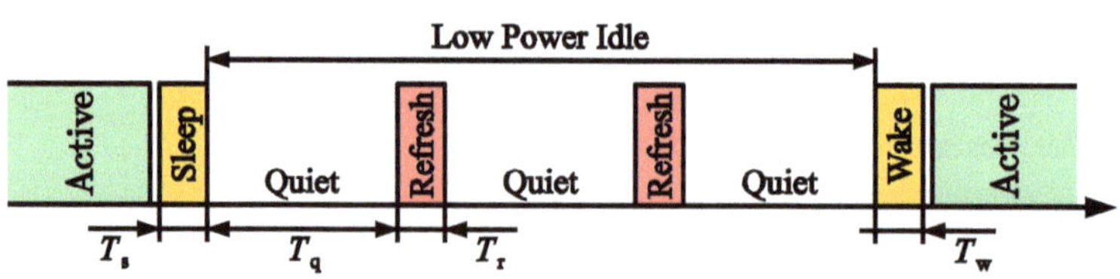

Fig. 1 Basic principle of energy efficient Ethernet, in accordance with [4]

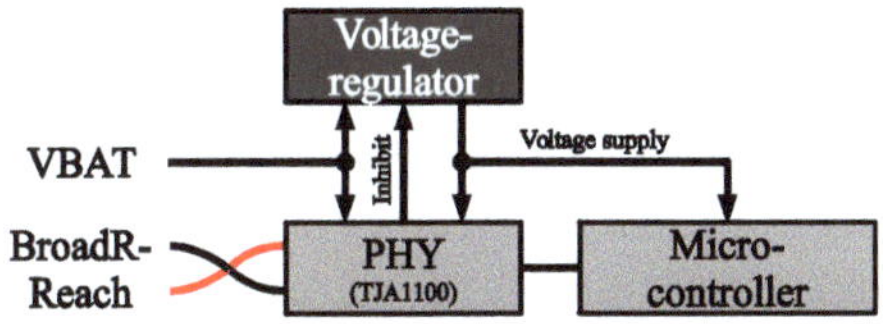

Fig. 2 Simplified block diagram of a low power sleep capable ECU

by the IEEE. It depends on the used Ethernet system and is 250 microseconds for 100 BASE-TX [4]. Energy Efficient Ethernet is currently specified for Ethernet systems operating between 100 Mbit/s and 10 Gbit/s [4]. For EEE operation an Ethernet system that uses a block coding scheme is required [6]. Therefore, BroadR-Reach is in principle suited for the adoption of EEE.

2.2 *Low Power Sleep Concept*

Low Power Sleep (LPS) is an energy optimization method developed by NXP. It's incorporated in the TJA1100 BroadR-Reach PHY. This technology allows hibernating Automotive Ethernet capable ECUs and waking them either via a local or remote wakeup. For the wakeup signalization, reserved BroadR-Reach code groups are utilized [7]. When a wake event is detected, the voltage-regulator is inhibited and the power supply to the ECU is restored. The simplified block diagram of a LPS capable ECU is illustrated in Fig. 2.

In contrast to EEE, the link is not maintained during the sleep mode. Thus a new linkup procedure is necessary once the PHY is woken up. Compared to EEE, this results in a considerably longer transition time. The targeted goal of NXP is 250 ms [7]. The LPS mode may therefore be utilized less frequently than the Low Power Idle mode of EEE. However, once a node enters the Low Power Sleep mode the energy saving is higher, since the complete ECU can be hibernated instead of just the PHY.

2.3 *Further Concepts*

Another technology that may be applied to Automotive Ethernet networks is Power over Ethernet (PoE) [8]. It provides the power supply and the data communication over the same channel. By turning off the power supply of idle links energy is preserved. A drawback however, is that PoE only allows remote wakeups. It may therefore not be suited for all applications. Nevertheless, it might be a viable solution for certain use cases, such as ADAS cameras, which shall start streaming data once they're activated. Due to the slow startup times (detection within 500 ms plus supply within 400 ms [9]) standard PoE is not suited for automotive applications. Another issue is that PoE is not specified for single twisted pair systems. This is currently being addressed by the 1-Pair Power over Data Lines (PoDL) study group. The resulting standard shall cover single twisted pair systems and also consider automotive requirements [10].

An energy optimization method with bi-directional wakeup capability is proposed in [11]. It uses the Simple Network Management Protocol (SNMP) for the network management. The wakeup signalization is performed using standardized link pulses, which are transmitted once a port is activated. These link pulses are detected by an energy detect module, which activates the power supply of the ECU [11].

The Low Frequency Wakeup (LFW) concept uses frequencies below 100 kHz for the wakeup signalization. Since these frequencies are usually filtered out, there's no interference with the data communication. A frequency detection module reacts to wake events and inhibits the power supply [12].

These three concepts are in a way comparable to LPS. They all break up the link and allow powering down the whole ECU. Since a new linkup is required, their mode switching is considerably slower than EEE but a higher saving potential is offered in either case. For the following considerations only LPS is taken into account. However, they may also be applied to another of these concepts, with only minor adjustments.

3 Analytical Assessment

In this section idealized estimations for the power consumption of Automotive Ethernet capable ECUs are made, using simplified analytical models. Hereby Energy Efficient Ethernet and Low Power Sleep shall be considered. While EEE may be used more frequently than LPS, the energy savings are not as big. This is illustrated in Fig. 3. Which method is more efficient in a given case, therefore depends on the traffic the ECU has to handle.

For all of the proposed considerations a data rate of 100 Mbit/s is assumed. Furthermore it's assumed the transition timing between the modes is ideal. During the transition from normal to sleep mode and vice versa, the power consumption is considered to be equal to the one in normal mode. Effects as shown in [2], that

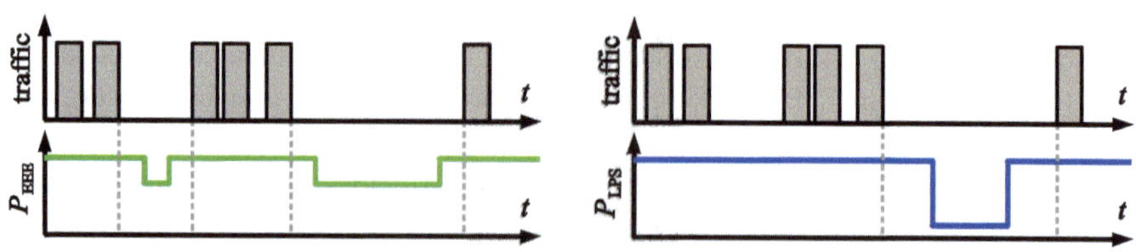

Fig. 3 Schematic comparison of EEE and LPS

ECUs may draw a higher current during initialization, thus reducing or even negating the efficiency of the applied technique, are not considered. Finally only a single Ethernet link is taken into account. Increased delays due to consecutive wakeups in multi-hop networks are not considered.

3.1 Model 1—Periodic Frames

This traffic model is proposed in [13]. It consists of single Ethernet frames, which are transmitted periodically. The size of the individual frames, as well as the time interval between two consecutive frames is constant. The power consumption $P_{PHY,EEE}$ of a PHY, which utilizes EEE, may be calculated according to Eq. 1 [13]. For link utilizations u below the threshold $u_{th,EEE}$, the power consumption follows a linear function. The link utilization threshold is the utilization for which it's no longer possible to enter the LPI mode, as the inter-frame duration is too short. It's dependent on the transition time T_{EEE}, the frame size s_{frame} and the data rate r_{data}. It can be calculated according to Eq. 2 [13].

$$P_{\mathrm{PHY,EEE}}(u) = \begin{cases} \frac{P_{\mathrm{PHY,max}} - P_{\mathrm{PHY,LPI}}}{u_{\mathrm{th,EEE}}} u + P_{\mathrm{PHY,LPI}}; & u < u_{\mathrm{th}} \\ P_{\mathrm{PHY,max}}; & u \geq u_{\mathrm{th}} \end{cases} \tag{1}$$

$$u_{\mathrm{th,EEE}} = \frac{s_{\mathrm{frame}}}{\left(T_{\mathrm{EEE}} + \frac{s_{\mathrm{frame}}}{r_{\mathrm{data}}}\right) r_{\mathrm{data}}} \tag{2}$$

When applying this traffic model, the effectiveness of EEE strongly depends on the frame size. When transmitting smaller frames, the duration between them has to be reduced in order to obtain the same link utilization. In Table 1 the link utilization threshold is given for various frame sizes. In [13] a calculation for asymmetric traffic is presented. This shall be disregarded, as BroadR-Reach only uses one wire pair. The LPI mode may therefore only be entered, when no communication takes place in either direction (same is the case for 1000BASE-T [6]). Unless otherwise noted, maximum sized frames (1530 bytes) are considered for the following calculations.

In order to compare EEE and LPS, Eq. 1 has to be adjusted. Since the higher saving potential of LPS stems from powering down the ECU, the power

Table 1 Link utilization threshold of EEE for different frame sizes

Frame size (bytes)	u_{th} (%)
72	2.25
500	13.79
1000	24.24
1530	32.87

Table 2 Assumed parameters

Description	Sign	Value
Power consumption of ECU (excl. NW components) ·	P_0	1000 mW
Power consumption of MAC	P_{MAC}	38 mW
Power consumption of PHY (normal mode)	$P_{PHY,max}$	300 mW
Power consumption of PHY (LPI mode)	$P_{PHY,LPI}$	30 mW
Power consumption of PHY (LPS mode)	$P_{PHY,LPS}$	1 mW
Transition time for EEE ($T_s + T_w\_sys\_tx$)	T_{EEE}	250 μs
Transition time for LPS	T_{LPS}	250 ms

consumption of the complete ECU and not just the PHY must be taken into account. The adjusted formulas are given in Eq. 3 and Eq. 4 respectively.

$$P_{ECU,EEE}(u) = \begin{cases} \frac{P_{PHY,max} - P_{PHY,EEE}}{u_{th,EEE}} u + (P_0 + P_{MAC} + P_{PHY,EEE}); & u < u_{th} \\ P_0 + P_{MAC} + P_{PHY,max}; & u \geq u_{th} \end{cases} \quad (3)$$

$$P_{ECU,LPS}(u) = \begin{cases} \frac{P_0 + P_{MAC} + P_{PHY,max} - P_{PHY,LPS}}{u_{th,LPS}} u + P_{PHY,LPS}; & u < u_{th} \\ P_0 + P_{MAC} + P_{PHY,max}; & u \geq u_{th} \end{cases} \quad (4)$$

The used parameters are explained in Table 2. The power consumption of the ECU P_0 contains all components, such as the microcontroller, sensors or actuators. In case the microcontroller also implements the Medium Access Controller (MAC) and the PHY, the values have to be adjusted accordingly. Unless otherwise stated, the values given in the Table 2 are used for the following calculations. It shall be noted that the used parameters are exemplary and may vary significantly depending on the application.

When considering maximum sized frames, the link utilization threshold of LPS is 0.0489 %, which is equivalent to 3.99 frames per second. For comparison, the according value is 2685.3 frames per second when using EEE. Besides the link utilization threshold, a second characteristic value is the turnover time t_{to}. It shall be defined as the period of time after which EEE and LPS are equally efficient. It is given in Eq. 5. The formula for the turnover time is independent of the used traffic model. It solely depends on the transition times and the respective power consumptions. Using the values, given in Table 2, t_{to} is equal to 313.20 ms. For the proposed traffic model, this is equivalent to a link utilization of 0.0391 % or the transmission of 3.19 frames per second.

$$t_{to} = \frac{T_{EEE}(P_{PHY,max} - P_{PHY,EEE}) - T_{LPS}(P_0 + P_{MAC} + P_{PHY,max} - P_{PHY,LPS})}{P_{PHY,LPS} - (P_0 + P_{MAC} + P_{PHY,EEE})} \quad (5)$$

In Fig. 4 the influence of P_0 on the turnover time t_{to} is illustrated. For low-power ECUs (P_0 is in the range of the power consumption of the network components) LPS becomes increasingly inefficient compared to EEE. The difference of the

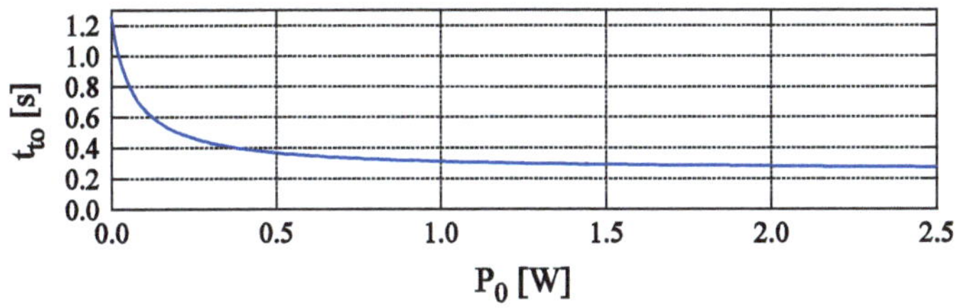

Fig. 4 Influence of the ECU power consumption on the turnover time

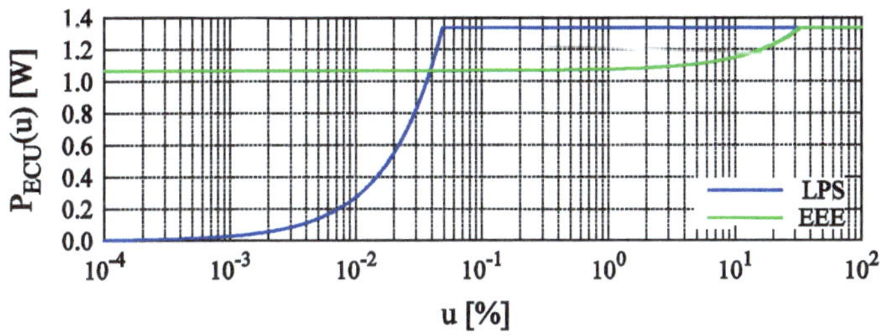

Fig. 5 Comparison of EEE and LPS for maximum sized frames

power consumptions in sleep mode is reduced, thus LPS requires a longer time to compensate its slower transition time. In contrast to this, the turnover time tends towards the transition time of LPS ($t_{to} \rightarrow T_{LPS}$) for high power ECUs. This means after entering the sleep mode, LPS is almost instantly more efficient than EEE.

In Fig. 5 the power consumption of an ECU as function of the link utilization is illustrated for both EEE and LPS. As basis for these calculations the values given in Table 2 are used. The figure shows the dependence of the ECU's power consumption on the link utilization. For link utilization smaller than u_{to}, which is 0.0391 %, LPS is more efficient. For $u_{to} < u < u_{th,EEE}$ the average power consumption is lower, when using EEE. For link utilizations above 32.87 % neither of the energy saving methods can be applied and the ECU requires the maximum power at all times.

Given these results, Energy Efficient Ethernet may present a viable alternative for Automotive Ethernet links that have to handle periodic traffic consisting of single frames, such as control data. Only for very low link utilizations the PHYs may be switched into the LPS mode, when this traffic model is applied.

3.2 *Model 2—Periodic Blocks*

The first proposed traffic model may not applicable to all automotive use cases. For example media streaming may require transmitting a large amount of data frames consecutively. In a first approximation this behavior shall be described by a second

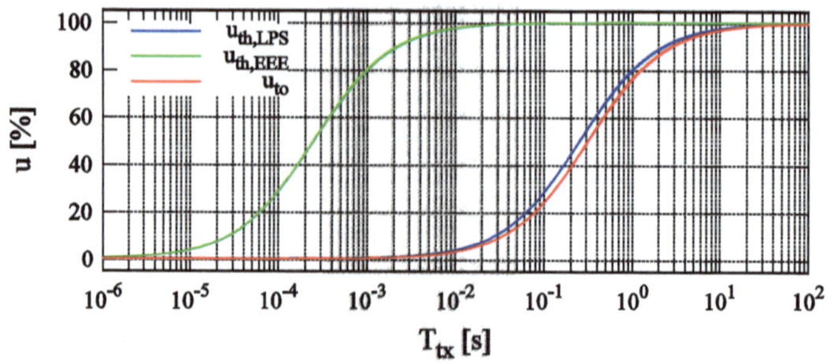

Fig. 6 Link utilization threshold as function of the block size

traffic model, which uses periodic blocks of traffic. This model is adopted from model 1. However, instead of single frames a block of consecutive frames is transmitted. There's no inter-frame gap between the frames that constitute a block. The size of the individual blocks, as well as the interval between them is constant. From an analytical point of view, the traffic blocks may be viewed as oversized frames. Therefore the basic behavior remains unchanged and Eqs. 3 and 4 may still be applied. Only the respective link utilization thresholds must be adjusted. The time necessary to transmit a single traffic block shall be denoted as T_{tx}. The link utilization threshold for this traffic model can be calculated according to Eq. 6, where $T_{tr,x}$ is the transition time of the respective energy saving methods.

$$u_{\mathrm{th,x}} = \frac{T_{tx}}{T_{tx} + T_{\mathrm{tr,x}}} \tag{6}$$

The influence of the block size on the utilization thresholds, as well as turnover utilization is illustrated in Fig. 6. With increasing block size the turnover point shifts towards higher utilizations, thus making it more likely that LPS may be used effectively.

3.3 *Model 3—Periodic Bursts*

Model 2 is a very rough approximation. Even when streaming data, it's unlikely that frames are transmitted without intermission for prolonged periods. For this reason, a third model consisting of periodic bursts is proposed. This model is periodic with each cycle consisting of two phases. During the idle time T_i no transmission takes place. During the burst, which has the duration T_b, even sized frames are transmitted with a constant interval between them. The behavior of the previous models may be reproduced using periodic bursts. When the idle time is set to $T_i = 0$, the behavior is equivalent to model 1. If the link utilization during the burst is set to 100 % this model is equivalent to model 2. The basic principle of this

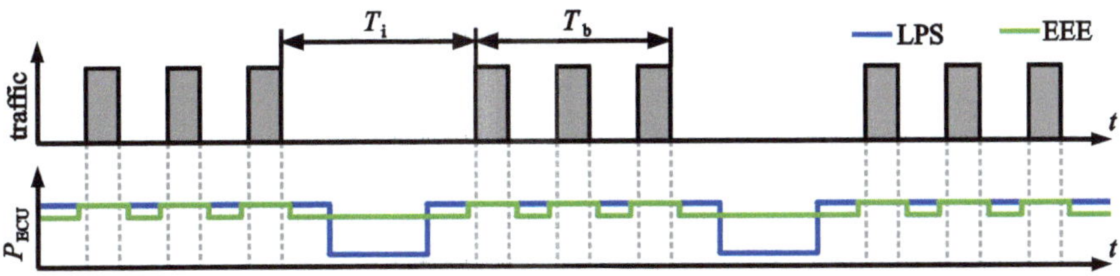

Fig. 7 Model 3—Periodic bursts

model is shown in Fig. 7. As illustrated, a constellation is conceivable, in which EEE may also be applied during the burst, whereas LPS is only possible between consecutive bursts. In general, the power consumption of an ECU using this traffic model can be expressed by Eq. 7, where P_i and P_b are the power consumptions during the idle period and the burst.

$$P_{ECU} = \frac{P_i T_i + P_b T_b}{T_i + T_b} \tag{7}$$

By definition, the link utilization is $u_i = 0$ during the idle period. However, a transition period, during which the power consumption is the same as in the normal mode (P_{max}) must also be considered. Therefore the power consumption is equal to P_{max} for the transition duration. For the remainder of the idle period ($T_i' = T_i - T_{tr,x}$) it is reduced. Considering that the power consumption follows a linear function during the burst, it may be expressed by Eq. 8, where m_x is the slope and y_x is the y-intercept.

$$P_{ECU,x} = \frac{m_x u_b T_b + y_x \left(T_i' + T_b\right) + T_{tr,x} P_{max}}{T_i + T_b} \tag{8}$$

When using the respective terms for the slope and y-intercept of EEE and LPS Eqs. 9 and 10 may be derived. These equations are only valid for link utilizations $u_b \leq u_{th}$, as well as for idle periods larger than the transition period ($T_i > T_{transition,x}$). The power consumption of an ECU in LPI mode is given by P_{EEE}.

$$P_{ECU,EEE} = \frac{\frac{P_{PHY,max} - P_{PHY,EEE}}{u_{th,EEE}} u_b T_b + P_{EEE}\left(T_i' + T_b\right) + P_{ECU,max} T_{EEE}}{T_i + T_b} \tag{9}$$

$$P_{ECU,LPS} = \frac{\frac{P_{ECU,max} - P_{PHY,LPS}}{u_{th,LPS}} u_b T_b + P_{PHY,LPS}\left(T_i' + T_b\right) + P_{ECU,max} T_{LPS}}{T_i + T_b} \tag{10}$$

As mentioned before, Eqs. 9 and 10 are only valid for $u_b \leq u_{th}$. For link utilizations above the respective threshold, neither of the energy optimization methods

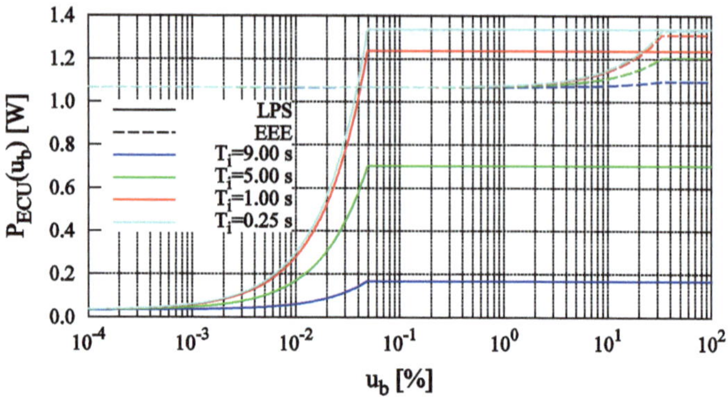

Fig. 8 Power consumption as function of the link utilization during the burst

may be utilized during the burst. In this case, Eqs. 11 and 12 have to be applied to calculate the power consumption of the ECU.

$$P_{\mathrm{ECU,EEE}} = \frac{P_{\max}(T_{\mathrm{b}} + T_{\mathrm{EEE}}) + P_{\mathrm{EEE}}T_{\mathrm{i}}'}{T_{\mathrm{i}} + \mathrm{T_b}} \tag{11}$$

$$P_{\mathrm{ECU,LPS}} = \frac{P_{\max}(T_{\mathrm{b}} + T_{\mathrm{LPS}}) + P_{\mathrm{PHY,LPS}}T_{\mathrm{i}}'}{T_{\mathrm{i}} + \mathrm{T_b}} \tag{12}$$

In Fig. 8 the power consumption of an ECU, as function of the link utilization during the burst, is illustrated for various combinations of the idle and burst durations. The overall duration of one cycle ($\mathrm{T_i} + \mathrm{T_b}$) is constant at a value of ten seconds in all cases. For certain constellations, especially when the idle period is very short, EEE may be more efficient. However, if the idle period is increased, LPS is far more efficient than EEE.

4 Analysis of Traffic Samples

The proposed analytical models are suited to obtain a first approximation of the power consumption. The used traffic models may reflect certain key characteristics of the traffic over an actual channel. However, it's not possible (with reasonable effort) to accurately model the actual traffic.

In order to do so, the analysis of the traffic samples may provide a better estimation. Therefore, an implementation based on an Octave script is proposed. As input parameters the frame size, as well as the starting time of the Ethernet frames are given to the script. This information may for example be extracted from a Wireshark trace. First, the script calculates the channel usage. Based on this information, the power states of EEE and LPS are calculated over the time. The results may then be

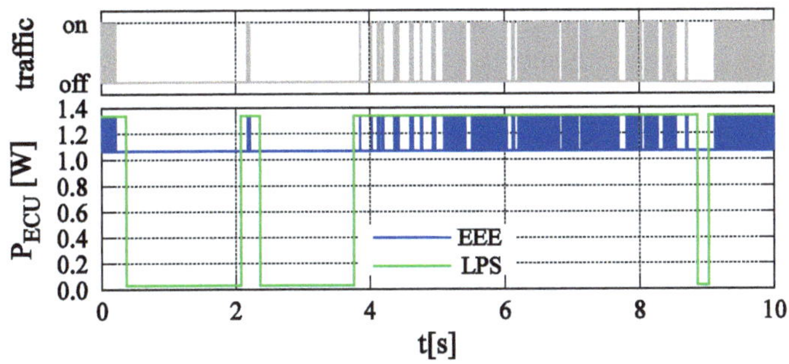

Fig. 9 Analysis of a traffic sample with the proposed script

visualized and the average power consumptions calculated. The same idealizations made for the analytical assessment, such as ideal timing, are also applied here. For the traffic sample illustrated in Fig. 9, the power saving is approximately 20 % when using EEE and 29 % when utilizing LPS.

5 Conclusion

The proposed idealized considerations suggest, that there's not a single energy optimization method that is suitable for all automotive applications. Instead the traffic, which shall be transmitted over an Automotive Ethernet link, strongly influences the efficiency of these techniques.

Low Power Sleep offers a high saving potential, as it allows hibernating the whole ECU. The downside of this is the need for a new linkup procedure after each wakeup, resulting in a longer transition time. Similar to Partial Networking on CAN buses, LPS is therefore best suited for the connection of ECUs, which aren't required at all times. Depending on the state of the vehicle, they may be powered down for prolonged period of times. However, the implementation of Automotive Ethernet requires the adoption of a switched network with point-to-point links. This also necessitates the introduction of switches. The links connecting these switches have to transmit the combined communication of multiple ECUs. Therefore these links will in average, have a higher utilization and shorter inter-frame gaps. Thus an efficient implementation of LPS is unlikely.

Energy Efficient Ethernet on the other hand, allows fast mode switching because the link integrity is maintained. EEE would therefore be an alternative for links, which may rarely utilize the LPS. Possible applications may therefore be the links between switches, or the links connecting ECUs, which are cyclically transmitting control data at short intervals. Another possible application for EEE may be low power ECUs. If the overall power consumption of an ECU is mainly dominated by the power consumption of the PHY, the shorter transition time of EEE may compensate the higher saving potential of LPS, independent of the traffic.

When considering the vehicle as a whole, the best energy efficiency may be gained, when applying a combination of multiple energy optimization methods. However, combining multiple approaches in the same system adds additional complexity to the network design.

References

1. Lim HT, Völker L, Herrscher D (2011) Challenges in a Future IP/Ethernet-based In-car Network for Real-time Applications. In: Proceedings of the 48th design automation conference, New York, pp 7–12
2. Grzemba A, Plaga S, Faschingbauer A, Scheer P, Fuchs M (2011) Partial networking. In: Proceedings of embedded world 2011, Nürnberg
3. Schmutzler C (2012) Hardwaregestützte Energieoptimierung von Elektrik/Elektronik-Architekturen durch adaptive Abschaltung von verteilten, eingebetteten Systemen. KIT Scientific Publishing, Karlsruhe
4. IEEE Standard for Ethernet (2012) Section Six, Institute of Electrical and Electronics Engineers, Standard IEEE 802.3-2012
5. Reviriego P, Christensen K, Rabanillo J, Maestro JA (2011) An initial evaluation of energy efficient Ethernet. IEEE Commun Lett 15(5):578–580
6. Spurgeon CE, Zimmerman J (2014) Ethernet: the definitive guide—Designing and managing local area networks, 2nd edn. O'Reilly & Associates, Sebastopol
7. Suermann T, Müller S (2014) Power saving in automotive Ethernet. In: Fischer-Wolfarth J, Meyer G (eds) Advanced microsystems for automotive applications 2014: smart systems for safe, sustainable and networked vehicles. Springer, Heidelberg, pp 93–100
8. Hank P, Suermann T, Müller S (2012) Automotive Ethernet, a holistic approach for a next generation in-vehicle networking standard. In: Meyer G (ed) Advanced microsystems for automotive applications 2012: smart systems for safe, sustainable and networked vehicles. Springer, Heidelberg, pp 79–90
9. Rech J (2007) Ethernet: Technologien und Protokolle für die Computervernetzung, 2nd edn. Heise, Hannover
10. Objectives for Power over Data Lines (2013) IEEE PoDL Study Group, Objectives Version 1.0
11. Balbierer N (2011) Energieeffizienz bei auf IP und Ethernet basierenden Fahrzeugnetzen. In *11. Ilmenauer TK-Manager Workshop: Technische Universität Ilmenau. 17. September 2010. Tagungsband*, Edited by Telekommunikations-Manager e.V., Universitätsverlag Ilmenau, Ilmenau
12. Seyler JR, Streichert T, Warkentin J, Spagele M, Glas M, Teich J (2014) A self-propagating wakeup mechanism for point-to-point networks with partial network support. In: Design, automation and test in Europe conference and exhibition (DATE) 2014, pp 1–6
13. Balbierer N, Waas T, Noebauer J, Seitz J (2011) Energy consumption of Ethernet compared to automotive bus networks. In: 2011 Proceedings of the ninth workshop on intelligent solutions in embedded systems (WISES), pp 61–66

Influence of the Design Parameters of Electric Vehicles in the Optimization of Energy Efficiency in Urban Routes

Alberto Fraile del Pozo, Emilio Larrodé Pellicer, Juan Bautista Arroyo García and Alberto Torné

Abstract The analysis and decision making on design, behavior and use of a prototype electric vehicle is the main focus of this paper. For this purpose, a prototype electric vehicle was modelled. The dimensional parameters needed to create the model were obtained by measurements, calculations and approximations. Subsequently, a route to be travelled by the vehicle was determined and a simulation of the vehicle on that route followed. Different modifications in the model were performed to compare their results with the original model through simulation in software Adams/Car. In this paper, we show the design criteria for electric vehicles to optimize engine power, weight, battery and vehicle performance according to the characteristics of typical urban routes.

Keywords Electric vehicles · Battery · Prototype · Simulation

1 Introduction

This paper is a collaboration between the Research Group on Sustainable Means of Transport and Systems (SMITS) of the University of Zaragoza and the Spanish company, Zytel Automotive S.L. (ZYTEL). The main goal is to use a software simulation tool such as Adams/Car for performing analysis and making decisions

A. Fraile del Pozo (✉) · E. Larrodé Pellicer · J.B. Arroyo García · A. Torné
Department of Mechanical Engineering, University of Zaragoza, C/María de luna s/n, 50018 Zaragoza, Spain
e-mail: afrailep@unizar.es

E. Larrodé Pellicer
e-mail: elarrode@unizar.es

J.B. Arroyo García
e-mail: jbarroyo@unizar.es

A. Torné
e-mail: atorne@unizar.es

T. Schulze et al. (eds.), *Advanced Microsystems for Automotive Applications 2015*, Lecture Notes in Mobility, DOI 10.1007/978-3-319-20855-8_16

on the design, behavior and use of a prototype electric vehicle. ZYTEL, as part of the development process and production of an electric vehicle called 'Gorila EV', decided to collaborate with SMITS for a study of the behavior and performance of the vehicle as well as to discuss possible improvements or changes in the architecture of it.

It was determined to tackle this problem by Computer Aided Engineering (CAE). Conducting an analysis of the vehicle by CAE allows increasing the speed and convenience when getting results and reducing costs incurred by testing with real prototypes.

In this study urban driving is mainly considered, where the vehicle trajectory is constrained by the infrastructure (road signs) and other vehicles (traffic). These constraints have a major impact on the vehicle trajectory, and therefore the formalization of the problem is different. In [1] an approach to define and evaluate the energy efficiency of a travel for an electric car driven in such conditions has been proposed.

The United Nations estimated that over 600 million people in urban areas worldwide were exposed to traffic-generated air pollution. Therefore, traffic related air pollution is drawing increasing concerns worldwide [2]. Electric vehicles (EV) are seen by many environmentally friendly groups and organizations as a potential solution to address the impact of transportation emissions in urban areas. Urban areas are also more suitable for the early adoption of electric vehicles due to the potential higher density of recharging stations.

1.1 Objectives

- Make a model of the Gorila EV as similar as possible to the real prototype and able to be simulated in software.
- Make a model of an urban route that can be used as a platform for the simulation of the vehicle model created and which is also capable of being used for the actual prototype for future trials and tests.
- Simulate the behavior of the vehicle model on the route and analyse the simulation results with an emphasis on those most important parameters.
- Suggest changes and possible improvements of the model and check its result and possible use in real life through simulation.

In order to achieve these objectives, collaboration of ZYTEL is needed primarily for the creation of a vehicle model that approaches reality as close as possible.

1.2 Overview of Electric Vehicles

Although a study of EV demand was published thirty years ago, most model-based academic studies of the likely uptake of Alternative Fuel Vehicles (AFVs) have been conducted in the last 10–20 years [3]. Recent issues of exhausting fossil fuels and global warming caused by internal combustion vehicles (ICV) have led to considerable efforts to develop EVs as environment-friendly vehicles utilizing electric energy sources [4, 5]. There is a large difference between the drivetrain system of EVs and ICVs. In contrast with conventional ICVs, the powertrain systems of EVs run electric motors through electrical energy stored in batteries [6]. EVs developed today don't differ so much from the more common ICV. The main components of the EV (rolling elements, chassis and bodywork) are essentially similar to those used by ICV; even if in the case of chassis and bodyworks, for example, efforts are made to reduce their weight in order to reduce energy consumption [7].

2 Modeling of Gorila EV's Architecture

2.1 Simulation Tool (Adams/Car)

Performing the analysis for a Gorila EV required a simulation software that would allow the construction of a model vehicle with high degree of detail and at the same time, the freedom to create our own analyses. For these two reasons, the Adams software was chosen, which among its modules has one specialized in the analysis of cars: Adams/Car.

With Adams/Car, one can quickly build and test functional virtual prototypes of complete vehicles and subsystems. Adams/Car makes the construction of vehicle models such as the assembly of the various subsystems that form it. Therefore, each subsystem should be constructed as closely as its real counterpart for subsequent assembly results in a complete vehicle similar to the real prototype and can be simulated accurately.

The Gorila EV model, whose behavior is simulated with Adams/Car, consists of the following subsystems: Wheels; Brakes; Front and rear suspension; Steering; Engine; Chassis. To approximate each subsystem to reality as close as possible, Adams/Car also allows configuration of various parameters (different for each of the subsystems). These parameters were obtained in three ways: provided by ZYTEL, calculated or approximated.

This study is based on previous work that consisted of a vehicle model made with a CAE program, SolidWorks (a design program for mechanical computer aided modeling), containing the representation of some of the above subsystems but could not be simulated. Adams/Car provides a library with simple models of all subsystems required for the simulation of the vehicle and will be used as the basis for creating the Gorila EV.

2.2 Subsystems Modeling

Since the powertrain system of an EV is a complex system with both the electrical and the mechanical sub-systems, it is necessary to model and analyse the powertrain systems of the EV in consideration of the two subsystems—to clearly address the vehicle dynamics. In particular, integrated modeling and analysis of the entire EV powertrain system is required to discover how a variety of vehicle dynamic characteristics are related with electrical quantities [8].

The configuration of the subsystems that form complete vehicle assembly has been determined from the data in each case. The necessary parameters were obtained through the company ZYTEL, taking the necessary measures or approximations, and using mathematical formulations [9].

Once, the modeling of all subsystems is completed, and then the data is assembled to form a complete vehicle and ready to be simulated and analyzed. In the case of the Gorilla, there is not an asymmetry with respect to the longitudinal load level exceeding 3 %. In Fig. 1, the center of gravity position of the Gorila EV and resulting assembly are shown.

The simulation, analysis and verification of the dynamic behavior of the Gorila EV were performed by scanning the vehicle model through a road created from an urban route that passes near the Campus 'Río Ebro' of the University of Zaragoza and whose modeling explained in the next section. The main reason for making a computer model of a real route as described is that it allows checking and implementing the simulations performed with the vehicle and comparing the results.

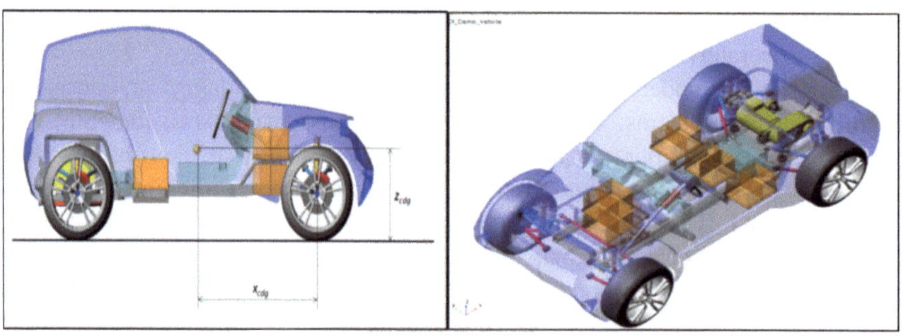

Fig. 1 Center of gravity position of the Gorilla EV, and assembly Gorila EV

3 Analysis and Results

This section describes first, the analysis carried out with the vehicle models and urban route created and it will serve as the basis for other simulations. It explains what it is and how such an analysis has been built. Next, it shows the subsystems that have to be studied because of their behavior, influence or qualities and what modifications have been carried out in respect to the basic analysis above. Finally, the results obtained with greater importance in these simulations are presented.

3.1 *Base Analysis*

In all analyses Adams/Car provided for assembly of complete vehicles has been chosen to carry out the simulation; and 3D Road allowed working with models of roads in three dimensions and has been chosen for its great flexibility to decide how one wants to travel on that model of road. To start the analysis, the following information was inputted: Vehicle assembly to analyse; Name to give to the analysis; Simulation Time (500 s); Number of steps (500); Simulation mode; Archive 3D road; Initial speed (10 km/h); Initial gear (1); Speed control; DCD file.

This analysis was performed by selecting the DCD file for control, since it is much more customizable than the introduction of the same acceleration for the entire track. In the case of Gorila EV, it was decided that the control of the vehicle speed would be based on the distance travelled. In the analysis of the Gorila EV, speeds that have been chosen are 20 km/h in the area corresponding to Campus 'Río Ebro' (both output as input), 30 km/h in roundabouts and 50 km/h on the straight areas.

3.2 *Analysis to Study the Behavior of the Subsystems*

Subsystems, whose behavior or influences on behavior of the vehicle under study were observed, were wheels, brakes, suspensions, chassis and engine. In addition to the above, an analysis of the vehicle with front wheel drive and analysis of vehicle energy consumption was also conducted. To carry out the study of the behavior of the wheels, there have been several analyses by varying the speed at which the vehicle travels the route. To do this, we simply amended the DCD file mentioned above by setting the vehicle traveling at roundabouts at higher speeds (40 and 50 km/h).

Furthermore, in order to analyse the performance of brakes, suspension and chassis, it was decided to change the load of the vehicle with an increase of the mass of the chassis (1250 and 1500 kg). This has forced a recalculation of the values of inertia and position of center of gravity of the vehicle. In the case of engine subsystem, its influence on vehicle behavior was evaluated by introducing new engines with different characteristics of weight, power and torque. Finally, the analysis of vehicle energy consumption required a previous estimate of weights of batteries needed to fulfil a certain number of cycles (a cycle is a complete route). This calculation then resulted in the creation of different vehicle models with different chassis masses depending on weights of batteries calculated. The types of batteries chosen to compare the vehicle energy consumption were lead-acid, nickel-cadmium and lithium-ion.

3.3 Results

3.3.1 Results of the Wheels Subsystem

It has been decided to focus the analysis of this subsystem on the rear wheels because they are the driving wheels and, hence, they are subjected to higher forces during the tour. First, the good qualities of the wheels regarding deformation and adhesion should be noted. The deformation of the wheels increases with higher vehicle speed and is more evident in the curved areas. However, the values always move in an environment no more than 15 mm in width, which means that the deformation is just over a centimeter despite the considerable increase in speed. Figure 2 shows the values of the radio of the wheels through the route (to the right wheel is similar).

Following the same line, the low longitudinal slips confirm the good adhesion of the wheel. This type of slip is measured as the percentage of the difference between vehicle speed and rotation speed of the wheel relative to vehicle speed. As mentioned, for the lateral sliding speed of 50 km/h, a greater difference was observed but it hardly attains a maximum value of 4 %.

However, although the results so far indicate a good performance of the wheels on the conditions simulated, Fig. 2 forces a complete change of idea. The Fig. 2 shows the angle of inclination of the wheels measured as the angle between the

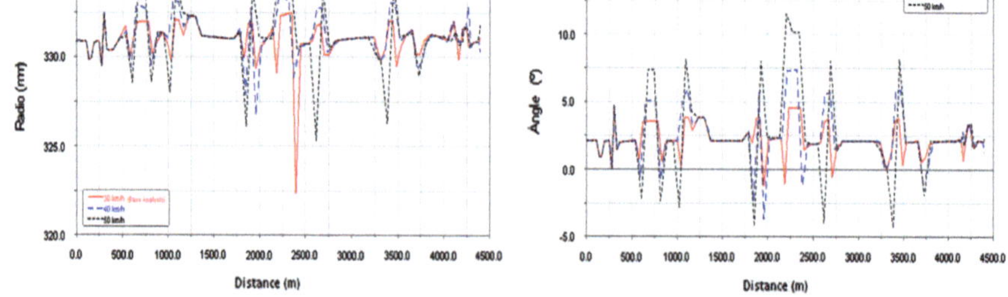

Fig. 2 Radio of the *left* rear wheel, and tilt angle of the *right* rear wheel

vertical plane and the plane of the wheel (in left wheel applies the same). For speed of 50 km/h (and almost 40 km/h) values were reached above 10° inclination which may damage not only the wheels but also parts of the suspension to which they are attached. Thus, it is clear that although the conditions of deformation and adhesion were good and the simulation was carried out in the program without problems, a solution was reached that was possible only from a mathematical point of view but was not from the point of view of real life. On the other hand, the positive conclusion that can be obtained was that the speed chosen for the base analysis (30 km/h) was adequate and did not jeopardize the vehicle.

3.3.2 Results of the Subsystem of Brakes

The subsystem of brakes is one of the least variety of output results provided in Adams/Car. From this variable, the program calculates other variables like the pressure in the brake line or the brake torque applied to the wheels. As can be observed, the brake demand significantly increases with the increasing of the weight of the vehicle both in the amount of braking and in the intensity, except for the last one in which the demand with the biggest weight is the lowest (due to the mistake in the convergence of the integrator to perform the calculations). In addition, it is also noteworthy that in spite of the considerable weight gain (up to 50 %) the brake demands range only 10 %. This is because the vehicle engine provides a very high brake torque and it is scarcely necessary to step on the pedal to adjust the vehicle speed to the specified.

3.3.3 Results of the Subsystems of Suspensions

The suspensions analysis has been centered on the rear suspension because it illustrates more clearly the behavior considering that it is subjected to greater forces. In essence, the behavior of the front suspension is the same as the rear apart from two exceptions: forces that work on the front suspension are minor and in moments of braking it is loaded and during accelerations, it is unloaded, unlike the suspension rear. Naturally, these forces increase when the vehicle accelerates, and decrease when the vehicle brakes in the same way, which forces are accentuated when a shift occurs.

3.3.4 Results of the Chassis Subsystems

The behavior of the chassis provides little surprises as increasing its weight and varies the position of the center of gravity and inertias of the vehicle. As would be expected, the warp angle, in general terms, increases as the mass of the chassis, without reaching dangerous limits. The pitch angle of the chassis is slightly greater with increasing weight, but the difference between one chassis weight and others is minimal and it is remained in normal values.

3.3.5 Results of the Engine Subsystems

The analysis of the influence of the engine in the vehicle has been made modeling different engines and creating new vehicle models. In this way, it had the original engine of 25 kW and three new engines of 30, 45 and 85 kW. These three engines are characterized by having more power, better engine inertia, less weight and a torque comparatively much higher, taking into account the difference in power of two of them which is not much bigger with respect to the original.

Finally, it is noteworthy that the energy consumption along a cycle (a full tour of the route) is less with any of the three new engines than with the original, despite the fact that the original has less power. In a single cycle, the most powerful engine is that which consumes less power, up to 5 % less than the base model engine. This is mainly due to three facts: weight (the weight of these engines is less than the original and the vehicle therefore weighs less and consumes less energy), inertia (the best inertia of these engines allows it to accelerate and brake using less energy) and torque [braking torque is much higher in these engines allowing them to recover more energy during braking (regenerative braking)].

3.3.6 Results of the Change of Rear-Wheel Drive to Front-Wheel Drive

This modification is simply to apply the torque to the front wheels instead of the rear wheels. Although at first sight, this may seem like a minor change, collected pitch angles and the deformation of the springs give us the idea that the vehicle does not behave exactly like the original. The vehicle with front-wheel drive brakes slightly stronger. A more detailed analysis shows that in some points in rear-wheel drive the engine uses more power to drive the same vehicle at the same speed than in front-wheel drive. This phenomenon is logical. Trying to simplify the explanation arguably is, that it is easier to pull an object (front-wheel drive) than to push (rear-wheel drive), so different powers are needed for it.

3.3.7 Results of Changing Batteries and Energy Consumption

In this analysis, a simple energy sizing of the vehicle is realized to roughly quantify the amount of kilograms of the battery that Gorila EV would need to fulfil a certain number of cycles. The method used for this calculation, in general terms, is based on a vehicle with a certain weight (including the weight of batteries to calculate) a certain autonomy and a particular type of batteries. From this, the weight of batteries that must be assembled on that vehicle to cover that distance is gained.

Table 1 Weight of batteries calculated and comparison of energies for some types of batteries

Batteries	Cycles	$M_{Batteries}$ (kg)	$M_{Vehicle}$ (kg)	$E_{Calculated}$ (Wh)	$E_{Simulation}$ (Wh)	Error (%)
Pb-Acid	10	130.61	845	2468.62	2640.03	6.94
Ni-Cd	10	87.93	800	2374.12	2536.94	6.85
Ion-Li	10	18.33	730	2227.11	2473.58	11
Pb-Acid	50	663.27	1375	17908.34	20184.72	12.71
Ni-Cd	50	394.07	1105	15073.23	16931.94	12.33
Ion-Li	50	87.93	800	11870.61	12684.72	6.86
Pb-Acid	100	2178.19	2890	67632.91	Error	–
Ni-Cd	100	912.59	1625	41066.88	66052.78	60
Ion-Li	100	192.19	905	25946.29	29361.11	13.16

The considered problem in this method is that the initial vehicle weight is not known since the weight of batteries to be assembled is unknown. Therefore, this calculation method should be applied iteratively until the appropriate value is obtained. In our case, the energy sizing compares the energy consumption of the vehicle with different weights for different types of batteries (lead-acid, nickel-cadmium and lithium-ion) to achieve a different number of cycles (10, 50 and 100 cycles), considering a cycle as the full tour of the route, which comes to about 4.4 km.

The result of the calculation gives weights of batteries reflected in Table 1.

To check whether calculations of the weight of batteries are correct, several models of vehicles have been built in Adams/Car, modifying the mass of the subsystem of the chassis to adjust it to the mass of the vehicle obtained. Once these models are constructed, they have been simulated in Adams/Car to obtain the consumed energy in a cycle. The number of cycles that must be covered can be compared regarding the results of the calculated energy previously and of the obtained energy by simulation (Table 1). Given that the weight of Pb-Acid batteries for 100 cycles is so high, that the simulation gives an error message and that the weight of Ni-Cd batteries for the same cycles have only mathematical sense, but not physical; then, the remainder of weights of batteries is maintained around of a 10 % of error with respect of the simulation, which is considered acceptable.

4 Study of Use Vehicle

In view of the results of the previous paragraph, it could be argued that the most influential factors in vehicle behavior are the engine, chassis and the type of batteries. The latter two are to be reduced to a simple change in the mass of the chassis. For this reason, it was decided to perform additional analysis creating some vehicle models in which the only difference between them was the engine or chassis (or both at once). Once these models were created, they have been simulated in Adams/Car and the results have been analyzed to see which would be the optimal use of each vehicle.

4.1 Creation of Vehicle Models

As it has been mentioned before, the factors most important for the vehicle behavior are the chassis load (factor A), the type of batteries (factor B) and the motor (factor M). First, to each factor two values from which to choose, one low and one high, must be given:

- A: load of the chassis
  A−: 200 kg. Simulating the weight of two occupants and some baggage, similar situation to the referred situation as base model in the Sect. 4.
  A+: 500 kg. Include the 200 kg mentioned and additionally an extra load in the vehicle of about 300 kg.
- B: type of batteries
  B−: Li-Ion. To achieve its specific energy about 5 times greater than of Pb-Acid batteries of the base model, it is considered a mass of 60 kg to take about the same amount of energy.
  B+: Pb-Acid. A mass of batteries is considered as the assembled mass in the base model, about 300 kg.
- M: engine
  M−: 25 kW. It is taken as low value, the engine of the base model since it has less power.
  M+: 85 kW. It is taken as high value, the most powerful engine of the mentioned engines provided by ZYTEL.

Before performing any combination of these factors, factors A and B are grouped into one. This is because both represent only the variation of the mass of the chassis subsystem. Thus, the number of factors and possible combinations is reduced to the half. The new factor will be the total load of the chassis and it is referred as factor 'C':

- C: Total load of the chassis
  C−: 260 kg. It includes passengers and Li-Ion batteries, so a lighter vehicle is obtained.
  C+: 800 kg. It includes passengers, load and Pb-Acid batteries, resulting in a much heavier vehicle.

Thus, the possible combinations are reduced to four: a vehicle with a very light or very heavy chassis that can carry a weak engine or a more powerful engine. After this, the four vehicle models are created in Adams/Car. For this purpose, first the two subsystems of the chassis with appropriate masses are created. It was not needed to do the same with engines because it had been modeled within the previous analyses. Having done this, these subsystems replace the corresponding in the base model of the vehicle and therefore four new assemblies of complete vehicles in Adams/Car are created.

4.2 Simulation Results of the New Models Created

The simulation carried out with the four created models was the same as in previous models. However, on the heavy vehicle, the weak engine is not able to reach the torque values of the most powerful. Regarding the brake demand, vehicles equipped with the most powerful engine do not need to use the brake, due to higher braking torque of the engine of 85 kW. In the case of the light vehicle, this difference is not so important as for the heavier model. regarding the accelerator pedal demand, it can be observed that vehicles with engine of 25 kW need to step on or let out the accelerator with more depth to reach values of torque that the motor of 85 kW reaches working in a smaller interval regardless of the vehicle weight (between 30 and 65 % of the demand). The developed powers of engines in the light vehicle are almost identical. On the heavy vehicle, the engine of 25 kW does often not provide as much power as the engine of 85 kW.

Finally, in the Fig. 3 the consumed energy at the end of the cycle is shown. For the case of the lighter vehicle, the differences are minimal, although the engine of 25 kW is the most energy efficient. On the other hand, in the case of the heavy vehicle the most powerful engine is the most appropriate, since the energy consumption is lower by far. This is due to the higher recovery of energy on the braking (because of its most braking torque).

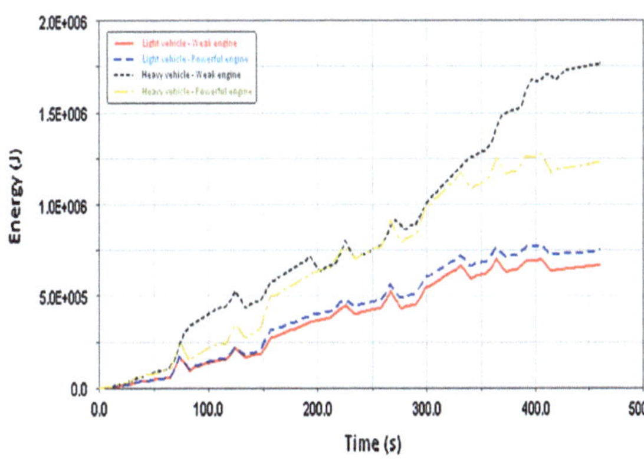

Fig. 3 Energy consumed by the chassis/engine combinations

4.3 Conclusions About Use

Given the weight of batteries for both vehicles, light and heavy, it can be assumed that both vehicles would be designed to circulate in city. However, what could be the use of the simulated vehicles, when going deeper into urban use? In the case of the vehicle with less weight, it could be equipped with a less powerful engine. As seen, this engine smoothly reaches the values of torque and power of the more powerful engine, vehicle braking is virtually the same and the power consumption is lower. Therefore, it would not be necessary to make expenditure with a more powerful engine that will not provide any further advantage. This vehicle would be perfect for people who, for example, go daily from home to work and from work to home, do some errand or go shopping. In the case of heavier vehicle, it should be equipped with the more powerful engine. It has already been shown that the 25 kW engine does not reach the required torque and power values and in addition the 85 kW engine would consume less energy through good regenerative braking system. This vehicle would be more suitable for fleets of delivery vehicles or carriers for services that have to carry some heavy equipment, ultimately, vehicles having to carry a freight and drive several kilometers a day.

References

1. Dib W, Chasse A, Moulin P, Sciarretta A et al (2013) Optimal energy management for an electric vehicle in eco-driving applications. Control Eng Pract 29:299–307 (Elsevier)
2. Zhou Y (2005) Modeling and simulation of hybrid electric vehicles. Ph.D. Dissertation. University of Science and Technology, Beijing
3. Shepherd S, Bonsall P, Harrison G (2012) Factors affecting future demand for electric vehicles: a model based study. Transp Policy 20:62–74
4. Butler K, Ehsani M, Kamath P (1999) A Matlab-based modeling and simulation package for electric and hybrid electric vehicle design. IEEE Trans Veh Technol 48:1770–1778
5. García P, Torreglosa J, Fernández L, Jurado F (2013) Control strategies for high-power electric vehicles powered by hydrogen fuel cell, battery and supercapacitor. Expert Syst Appl 40: 4791–4804
6. Park G, Lee S, Jin S, Kwak S (2014) Integrated modeling and analysis of dynamics for electric vehicle powertrains. Expert Syst Appl 41:2595–2607
7. Chang D, Morlok E (2005) Vehicle speed profiles to minimize work and fuel consumption. J Transp Eng 131:173
8. Guzzella L, Sciarretta A (2007) Vehicle propulsion systems introduction to modeling and optimization. Springer, Berlin
9. Gao D, Mi C, Emadi A (2007) Modeling and simulation of electric and hybrid vehicles. Proc IEEE 95:729–745

Design and Systems Integration in the Electrification of an Electric Vehicle for Long Distance Travel. Hierarchical Multi-criteria Analysis for Designing the Vehicle Architecture

Emilio Larrodé Pellicer, Alberto Fraile del Pozo and Juan Bautista Arroyo García

Abstract This paper describes part of the development of racing electric vehicle "Zytel-Zero" which was developed in the laboratories of the I3A (Aragón Institute for Engineering Research). The main objective of the paper is the description of the integration of all systems and components of this electric vehicle, designing and building the subsystems for correct operation, and commissioning and preparation of the control system. Special focus has been laid on the control systems, especially the batteries and several sensors of vehicle itself. In this work the design and integration of systems required in the electrification of a long-range vehicle is described allowing it to be used as a testbed for monitoring battery consumption as a function of the scenarios and the number of batteries on-board. In this paper the use of the methodology Analytic Hierarchy Process (AHP) is presented, as a technique for multi-criteria decision aid to decision making in the process of the electrification of a vehicle.

Keywords E-vehicles · Battery · Electrification · Analytic hierarchy process (AHP)

E. Larrodé Pellicer (✉) · A. Fraile del Pozo · J.B. Arroyo García
Department of Mechanical Engineering, University of Zaragoza, C/María de luna s/n, 50018 Zaragoza, Spain
e-mail: elarrode@unizar.es

A. Fraile del Pozo
e-mail: afrailep@unizar.es

J.B. Arroyo García
e-mail: jbarroyo@unizar.es

T. Schulze et al. (eds.), *Advanced Microsystems for Automotive Applications 2015*, Lecture Notes in Mobility, DOI 10.1007/978-3-319-20855-8_17

1 Introduction

Initially, this project was carried out in the company ZYTEL Automoción S.L. in collaboration with the University of Zaragoza. ZYTEL is a Spanish company dedicated to the electrification of vehicles and electric vehicle manufacturing. As part of the development process of their electric vehicles and promoting the brand, ZYTEL has decided to participate in the ZERO RACE, where the electric vehicle will go around the world with zero CO_2 emissions.

The racing electric vehicle "Zytel-Zero" was developed in the laboratories of the I3A (Aragón Institute for Engineering Research).

A Reciprocating Internal Combustion Engine (RICE) powered vehicle was transformed for a high-performance autonomy electric vehicle. For this project, an Sport Utility Vehicle (SUV) was chosen due to the ease of structural changes (it has a separate frame), its cargo space for a large number of batteries, its lightweight and ability to cope with intensive practice for all types of soils.

The sizing of the vehicle was made keeping in mind the demands of career and regulations of the European Union [1–4], highlighting the following:

- The estimated daily travel distance is in the range of 400–450 km. The vehicles must have a range of at least 225 km fully charged until the first refueling of outage batteries, traveling at an average speed of 80 km/h. Following a reload of 4 h, it should travel another 200 km.
- The vehicles will be charged during lunch and overnight stops. 220 V outlets and 16A will be made available, and occasionally even a 380 V in Europe. A vehicle can have up to 3 power outlets.

This research develops a technology evaluation model, which incorporates technological factors and market criteria to facilitate decision making on designing the electric vehicle architecture. Using analytic hierarchy process (AHP), the technology evaluation model was quantified by experts in these technologies. Expert quantification ranks the technological alternatives in order of technological performances. This method helped the developments within this project and it will deploy strategies for the future.

2 Objectives

The fundamental objective of this project is the development of the vehicle and control system for proper operation. The project started very complex, since it was working comprehensively around the vehicle, any small change in one part of the vehicle forced to make other changes in the control systems, and even in the very physiology of the vehicle. Every little change or variable must be studied from a global perspective.

For clarity, it was decided to divide the paper into 5 blocks corresponding to the work done on the vehicle: the drive system, the power system, the control system, the vehicle structure and auxiliary systems.

- Drive System: This system refers to everything related to the drive unit (controller + motor) of the vehicle. Being mounted purely mechanical, this project will focus on its electrical wiring, control and communication between the system and the user of the vehicle by means of different sensors and signals.
- Feed system: The vehicle has more than 100 lithium iron phosphate ($LiFePO_4$) batteries with a total voltage that can reach 420 V and up to 500 A. Right control and monitoring of loading and discharging of the batteries as well as of their state at all times of operation of the vehicle is therefore a crucial point.
- Control system: For proper interaction between the user and the vehicle, control and monitoring of the different variables of the system is needed. The car has been equipped with a robot that will perform these functions and control the electrical circuits of the vehicle, providing a useful and attractive user interface.
- Vehicle structure: When working in the various vehicle systems, necessary modifications have repeatedly been made to the bodywork and the internal structure of the vehicle.
- Auxiliary systems: In addition to all the changes already mentioned, it was necessary to evaluate, test and make the necessary changes in the auxiliary vehicle systems that have been previously in the car as well as in those newly added.

3 Planning and Organization

3.1 *Work Scheme*

First, we worked on testing the initial state of the vehicle and studied existing regulations for this type of vehicles, especially regulations that are the most restrictive. After studying both, its status and the obligations imposed by regulatory roadmap, requirements and elements to be mounted inside the vehicle were marked. After this identification, the next step is to find solutions to these problems and needs. These solutions often require having to go to the market to search for the components that best fit with their subsequent market research, and adapting these solutions to the products found.

The next step is more practical: to implement the solutions adapted to the vehicle. This section includes the physical placement of different elements, the electrical wiring, and the programming, amongst others.

Finally, there is the realization that the solutions adopted and implemented meet the initial objectives and needs. The complexity of this process is that the vehicle must be treated in a comprehensive manner, the amount of variables and internal components is very high, and the interplay between each of them is complex. That is, a small change in one-component or vehicle systems will cause changes in the design, siting or installation of others.

3.2 Global Stages

There is a strong interdependence between the different systems described, so it has been necessary to work on the five systems simultaneously, keeping an overview of the vehicle, with the problems that it entails (Fig. 1).

When working on the drive system, it has been necessary to develop the system soft-start connecting it to the power system, so it was necessary to have all the wiring prepared. Further, to modify the motor variables and correct operation it was necessary to bring prepare the control system accordingly. The power system, as already mentioned, must be controlled for proper and safe operation through the

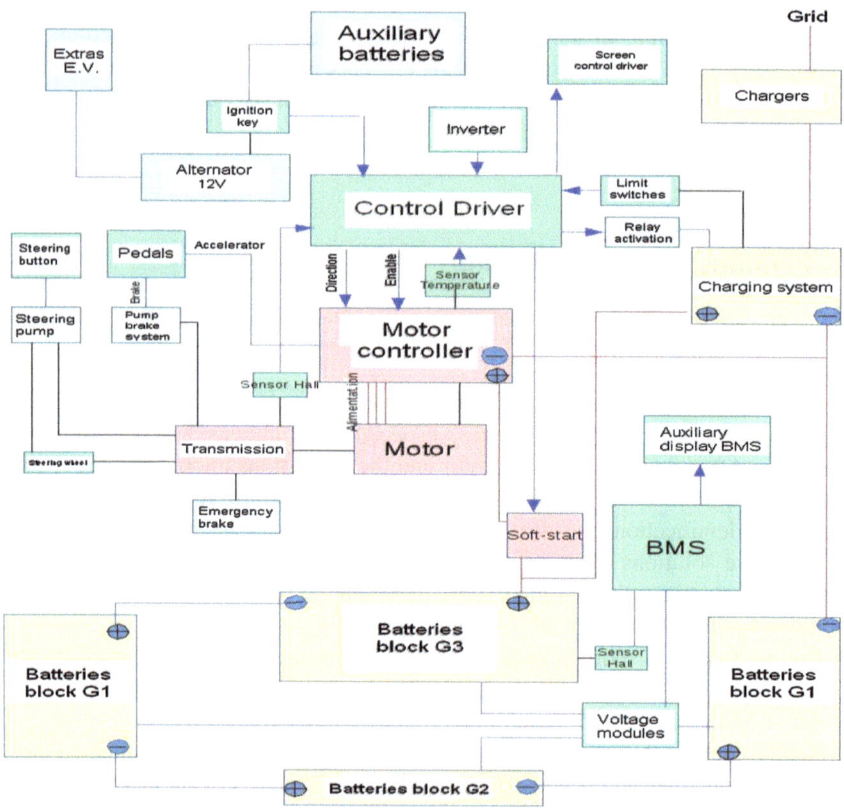

Fig. 1 Vehicle scheme

Battery Management System (BMS), so once assembled it should take their variables to the control system. However, for correct assembly it was to make many changes in the structure of the vehicle, as discussed above.

In turn, for the assembly of different devices for the vehicle control system also changes in the physical structure of the vehicle were necessary, which should be consistent with those made to adjust the fueling correctly.

Finally, changes in the auxiliary systems should also affect the power control systems so the monitored variables would behave at an optimal level.

4 Drive System

In the next section the results and characteristics of the work done in this project on the engine and the vehicle drive group is detailed. The vehicle is equipped with a brushless DC motor with permanent magnets, due to its high torque, high speeds, high acceleration, high reliability and speed-torque characteristic linear [5]. Apart from the electrical connections described above, communication with the user or driver is necessary. The driver must pass the tractor group (controller + motor), their indications of movement, speed, direction, etc. This information is transmitted to the controller via an Amphenol-19 connector [6].

The vehicle's engine needs a cooling system. This circuit consists of the cooling water pump (in charge of refrigerant flow), the main radiator, expansion tank (set to balance the pressures of the refrigerant circuit and storing), and the sleeves and would run the engine and its controller.

Inverters architecture requires the use of a buffer or energy storage capacitors (lighter and cheaper than the coils). Capacitors that provide a fast response due to their low internal resistance tend to be loaded with very high current peaks. To reduce this intensity, the solution adopted is the introduction of a circuit current limiting during charging of the capacitor (see Fig. 2). This system is called soft-start charge circuit and is composed of two contactors, two fuses and power resistor connected as shown in Fig. 2.

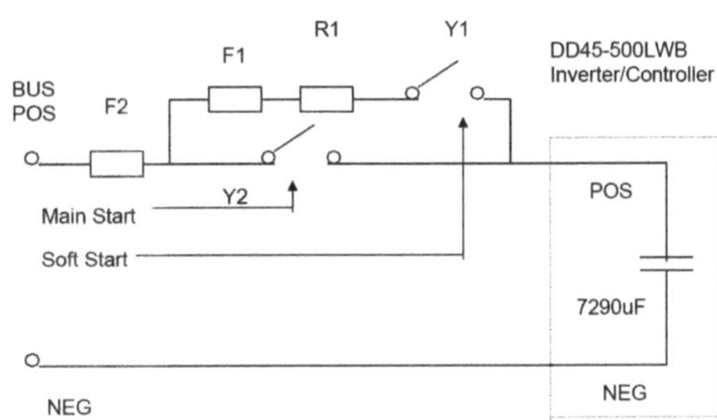

Fig. 2 Circuit soft-start connection

The power resistor connected in series with the inverter input limits the peak intensity or, equally, extending the charging time of the smoothing capacitor load chart.

5 Feed System

Today it is difficult to store large amounts of energy, especially in the form of electricity. Electric energy storage systems are expensive and heavy, so a previous calculation of electric power that will be used to achieve the vehicle to reach established targets is essential [7]. Through the determination of the equations governing the forces opposing the advance of the vehicle and the initial conditions, an algorithm will be set that will result in planning the necessary energy that would be needed to store. This process is called power sizing. The main function of power sizing process or Calculation Model of the Power Sizing (CMPS) is to determine the energy required to make a particular vehicle perform a certain number of drive cycles. Parallel it is used for various functions such as calculation of the engine power required, the maximum vehicle speed, and the acceleration, the selection of the gear ratio and the obtaining of the vehicle's curves among others [8, 9]. The energy required for the traction of the vehicle is set by the simultaneous sum of the resistance to advance, the speed and the time elapsed during the movement. Resistance to advance is caused by the friction of the tyres with ground, the friction with the air, inertial resistance and strength, positive or negative, produced in the ascent or descent of terrain slopes. By fixing the methodology for calculating the resistance to the advance of the vehicle, it will study consumption cycles in the regulation-101 [4].

5.1 Driving Cycles

A drive cycle graphically represents the vehicle's speed versus time during a sup-posed route. Throughout, a cycle one can distinguish four driving modes: acceleration, deceleration, constant speed and stop. The standard driving cycles [cycle ECE-15 and cycle EUDC (Extra-Urban Driving Cycle)] have been developed to simulate a conduction that is representative of urban and interurban driving. These cycles are used to evaluate the use of vehicles on test benches simulating real operating conditions. The cycle NEDC (New European Driving Cycle) is used in the European Union Regulation 101 [4] to certify passenger vehicles and light trucks. It is composed of four urban cycles ECE-15 and a cycle EUDC thus simulating real conditions daily. It has a duration of 1200 s at an average speed of 33.60 km/h and a total distance of 11,022 m. The result shows figures that are difficult to interpret at first sight therefore are transformed to more conventional

Table 1 Batteries possible to incorporate

Batteries	Lead-acid	Lithium	Sodium
Energy density (Wh/kg)	45	115	120
Weight_necessary (kg)	1506.96	589.68	565.11
Energy density (Wh/kg)	45	115	120

units, as it is consumption in kWh/100 km, so for this case get consumption per cycle of 17.23 kWh/100 km.

5.2 *Estimated Battery Weight*

Following the selection criteria through the regulatory requirements of the project, the weight of batteries must be set as a determining factor when selecting the vehicle and batteries. To do so, according to regulations of the racing committee the first stage of 225 km with a maximum recharge of 4 h and then another stage of 225 km must be tackled.

Concluding the first phase of power sizing, without knowing what the structural components of the project will be, it can be said that, looking at Table 1, the first selection is either lithium or sodium batteries. A light vehicle, large load capacity, a drive system as light as possible and high efficiency will be selected. It can be seen that weight is a key factor in the development of the project, mainly due to two reasons; one of them is the high percentage of mass, which will provide the batteries and the second the increased consumption by the weight, which would require a greater number of batteries, converting the calculation in an iterative process

5.3 *Choice of Batteries*

The choice of the batteries in the previous stage was conducted according to criteria of weight, maximum discharge, and engine power and cycle life of the batteries. Finally, the selected battery was the manufacturer SE180AHA Sky Energy. These were lithium batteries 3.3 VDC, a capacity of 180 Ah and a weight of 5.6 kg. The vehicle has been equipped with 106 batteries, making a total of 349.8 VDC, with a minimum voltage of 212 VDC. The total weight of the batteries is 593.6 kg. However, due to its placement, batteries may be added or removed according to current needs. The vehicle was sized so it can have a range of up to 500 km. Therefore, the final amount will depend on the battery needs studies final land where this prototype will be used. Because the vehicle is intended to be a self-pointer, the number of batteries is very high. Therefore, placement is not trivial. Their location of batteries in the vehicle were split in four blocks or coffers, two side skirts (G1), a rear flap (G2) and the inner drawer, in the area inside the vehicle (G3). The battery control system or BMS will give information on the voltage level of the batteries. When the system reaches a minimum value for the engine-operating load of the batteries is required. For this, it has two clips that go on the back of the vehicle, one single-phase and another three-phase.

6 Control System

One of the main areas of work in the development of this project has been the design of the control system. The control system is divided into two parts: first the controller, which controls the state variables of the vehicle and on the other, the BMS or battery management system to control the level and condition thereof. Given the great importance of vehicle power system, with its large number of batteries connected in series, it is necessary to introduce a system to monitor and control the status of each battery.

The BMS system allows for a real time reading of the voltage of each battery and the temperature of each set with ten batteries. It also allows the display of the current flowing in or out of the battery pack. The touch screen affixed to the car dashboard will show the total voltage of the batteries, the intensity that runs the stored energy resistance, higher voltage battery, low voltage battery and probe which has the highest temperature.

7 Vehicle Structure

Throughout the development of the project, it has been necessary to make changes to the structure of the vehicle to adapt to new systems coupled thereto. These changes are principally due to the need for cables between different parts of the vehicle and the need to add new devices to the dashboard or the area near the vehicle driver. Different systems have been installed in the vehicle for control and monitoring of important variables. Such monitoring must be visible to the user of the vehicle, so it is vital to have this information in an area easily accessible to the driver of the vehicle. This information is reflected in two different screens: the control driver and BMS or control system batteries.

8 Auxiliary System

Systems that integrate electric motors into a vehicle equipped with an internal combustion engine vary due to amending or removing some of its components. In addition, the operation of the electric traction vehicle is necessary to introduce other new systems. Brake system, steering system, auxiliary battery system, electric power system and the system of traction's batteries—Battery Management System (BMS) are new systems that will be incorporated or modified but which have not been detailed descriptively.

In addition to the systems already described, the operation was tested and the appropriate additional changes were made to the rest of the vehicle: lights, windows, dashboard, lights/turn signals, wiper and washer lever, horn, heating system, window defroster, etc.

Another important point of this work was the determination of the maximum temperature attainable by the vehicle batteries to ensure the safety of both the occupants and the battery system, designing, if necessary, the required cooling system. The calculations necessary for this were done by making a series of assumptions about the system under study focusing on the generation of heat inside the batteries and their interaction with the surrounding elements concerning the heat transfer, which will take place by conduction and convection. For this task it theoretical calculations, instrumental calculations (thermocouple) and calculations using a thermal imager were compared [10, 11].

9 Hierarchical Multi-criteria Analysis

Multi-attribute value theory has been used in decision analysis, which is a way of systematically analyzing and modeling the preferences among the alternatives [12]. Analytic hierarchy process (AHP) method structures the decision problem in levels, which correspond to one understanding of the situation: goals, criteria, sub-criteria, and alternatives. Its main steps are the following:

- First the relative importance of the criteria has to be defined on the basis of pairwise comparison assessments using a scale as shown in Table 2. For instance, if sub-criterion *Weight* is judged to be "Strongly Preferred" to sub-criterion *Durability*, a score of 5 is given; the opposite judgement is expressed with the reciprocal value (1/5). The results of pairwise comparisons are organized in a square matrix (A) having a number of rows (and columns) equal to the number of criteria to be compared (for instance n). The a_{ij} value of the matrix represents the relative importance of the ith criterion relative to the jth criterion. It is a matter of fact that only $(n-1)!$ Of the n^2 values of the matrix are independent and hence, only these values have to be obtained directly from decision makers. Indeed the other ones are automatically generated considering both reflexivity (if $i = j$ then $a_{ij} = 1$) and reciprocity ($a_{ij} = 1/a_{ji}$). The relative values forming the matrix (A) are translated in absolute priority weightings on the basis of Saaty's eigenvector procedure: $A \cdot W_j = k \cdot W_j$, where: W_j represents the vector of the absolute values of the importance weightings and k is the highest of the eigenvalues of the matrix A.

Table 2 Measurement scale for the AHP Method

Verbal judgement	Degree of preference
Equally preferred	1
Moderately preferred	3
Strongly preferred	5
Very strongly preferred	7
Extremely preferred	9

- Then the alternatives have to be assessed on a pairwise basis with respect to the criteria using the same procedure as described in step 1. The output achieved is the absolute rating of the alternatives with respect to all criteria (A_{ij}).
- Averaging the absolute rating with respect to each criterion and sub criterion with the corresponding absolute importance weightings. It is possible to calculate the overall suitability ratings (A_j) of the alternative investments. The AHP-based ranking was systematically obtained using Expert Choice software.

The AHP model is depicted in the Fig. 3 and several graphics of the hierarchical analysis are shown in the Fig. 4.

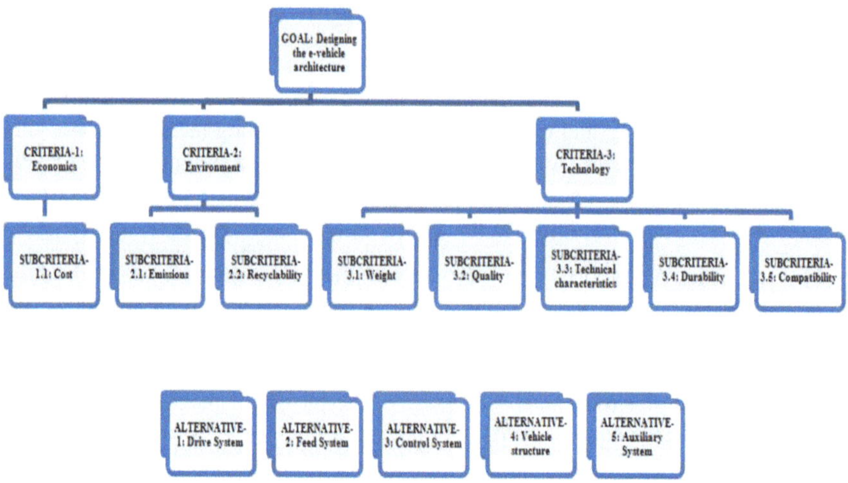

Fig. 3 The analytic hierarchy model

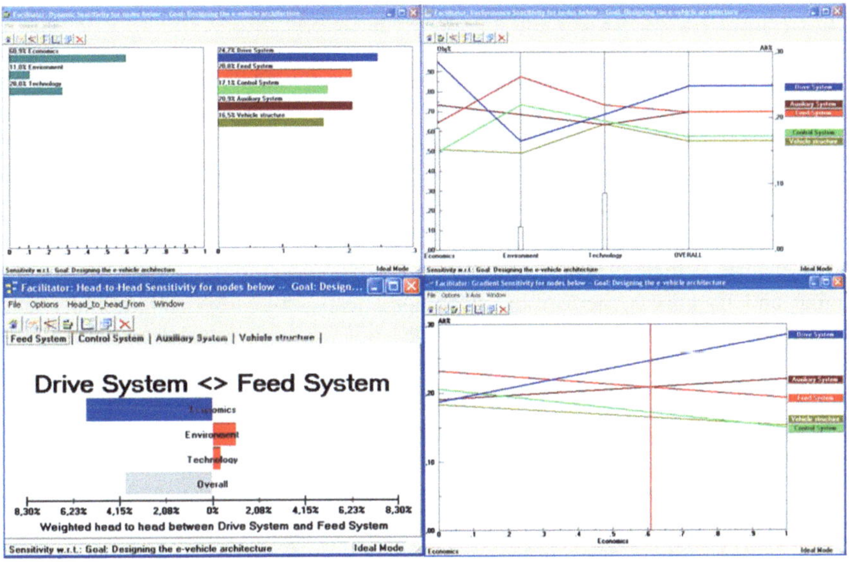

Fig. 4 Graphics of the hierarchical analysis results

10 Summary/Conclusions

The project began with the vehicle transformed by reciprocating internal combustion engine to an electric vehicle, pending the engineering integration, all wiring, electrical design and control. In developing this project, the objectives of setting up the operation and control system have been met and a vehicle was prepared ready for driving. Figure 5 shows all the added and designed electrical and control operation of the motor connections as well as all the wiring in the power system

Fig. 5 Final vehicle

with their corresponding protections and security features according to European legislation.

A monitoring system was also installed that can monitor the most important variables of the system and see the detailed battery status at all times, giving the user detailed vehicle information at all times.

Therefore, the electric vehicle is fully equipped and prepared for the requirements and begins the first tests. However, there are multiple options below, some of which are already being carried out in order to increase the degree of optimization of the electric vehicle. These recommendations should emphasize the study of the engine and its curves, performing behavioral tests, both tests rollers and actual performance. In addition, this is a prototype, so it would be interesting to study the number of batteries /power required for different land requirements and developing a database with all this information. Another field of work, which is already being tackled is the vehicle energy study, studying heat dissipation and means for cooling the battery system.

The vehicle on which the work was carried out was a test vehicle, on which various modifications over time were performed and none of the elements that have been installed in the first place necessarily corresponded to the final version.

Using the AHP model was helpful in choosing the components to install in the vehicle and to rank the alternatives construction.

References

1. Directive 2007/46/EC (2007) Parliament and European Council
2. Directive 70/156/EEC (1970) European Council
3. Regulation 10 and Regulation 100. Directive 2007/46/EC
4. Regulation 101. Directive 2007/46/EC
5. UQM Motor Systems User Manual
6. Fink A, Beaty H (2007) Standard handbook for electrical engineering. McGraw-Hill, New York
7. Fraile A, Sánchez S, Larrodé E. (2013) Electric vehicle preparation for vehicle performances analysis and battery evaluation in normal operation conditions. In: Advanced microsystems for automotive applications - smart systems for safe and green vehicles. Edit. Springer International, pp 233–243.
8. Linden D (2010) Handbook of batteries. McGraw-Hill, New York
9. Affanni A, Bellini A, Franceschini G, Guglielmi P et al (2005) Battery choice and management for new-generation of electric vehicles. In: IEEE transactions on industrial electronics
10. Wu M, Liu K, Wang Y, Wan C (2002) Heat dissipation design for lithium-ion batteries. J Power Sources 109:160–166
11. Sabbah R, Kizilel R, Selman J, Al-Hallaj S (2008) Active (air-cooled) vs. passive (phase change material) thermal management of high power lithium-ion packs: Limitation of temperature rise and uniformity of temperature distribution. J Power Sources 182:630–638
12. Ho C, Huang Y (2014) Evaluation of electric vehicle power technologies. In: Proceedings of PICMET'14: infrastructure and service integration

Opportunities for European SMEs in Global Electric Vehicle Supply Chains in Europe and Beyond

Frauke Bierau, Pietro Perlo, Beate Müller, Arrate Alonso Gomez, Thierry Coosemans and Gereon Meyer

Abstract Smart electric mobility is being regarded one of the major paths to sustainable mobility at least for urban areas. Governments, societies and industry in many countries already pursue this path. Some promising markets for electric vehicles (EVs) are evolving and many more emerging. This development provides opportunities for European small and medium sized enterprises (SMEs) in global EV supply chains, especially since electric mobility requiring new technologies, services and business models, changes the traditional automotive supply chain and opens it for new players. The EU-funded Coordination Action GO4SEM is aimed at spreading awareness of global market trends and opportunities and on investigating matching innovation capabilities of European SMEs. This analysis results in strategic advice towards SMEs, but also policy recommendations towards public authorities for strengthening SMEs in this endeavor are derived.

Keywords Electric mobility · Global markets · Automotive supply chain · Innovation · SMEs

F. Bierau (✉) · B. Müller · G. Meyer
VDI-VDE Innovation + Technik GmbH, Steinplatz 1, 10623 Berlin, Germany
e-mail: frauke.bierau@vdivde-it.de

B. Müller
e-mail: beate.mueller@vdivde-it.de

G. Meyer
e-mail: gereon.meyer@vdivde-it.de

P. Perlo
Interactive Fully Electrical Vehicles (IFEVS), Turin, Italy
e-mail: pietro.perlo@ifevs.com

A.A. Gomez · T. Coosemans
Mobility, Logistics and Automotive Technology Research Centre (MOBI), Electrical Engineering and Energy Technology Department (ETEC), Vrije Universiteit Brussel Mobility, Pleinlaan 2, 1050 Brussels, Belgium
e-mail: arrate.alonso.gomez@vub.ac.be

T. Coosemans
e-mail: thierry.coosemans@vub.ac.be

T. Schulze et al. (eds.), *Advanced Microsystems for Automotive Applications 2015*, Lecture Notes in Mobility, DOI 10.1007/978-3-319-20855-8_18

1 Introduction

Around the globe there is an increasing demand for smart, environment-friendly vehicles integrated in a sustainable and advanced mobility system. The World Business Council for Sustainable Development (WBCSD) defines sustainable mobility as: "the ability to meet the needs of society to move freely, gain access, communicate, trade and establish relationships without sacrificing other essential human or ecological values today or in the future". Smart and innovative solutions can fill the gap between the fast growing urban population and existing transport infrastructures on one side, and the need for safer, greener and more convenient mobility solutions on the other side. Electric mobility is the prime candidate to meet this demand.

Electric mobility has been introduced in European and international markets and the majority of OEMs are engaging in this development, many already have vehicle models on the market. The globally rising need for electric vehicles presents promising opportunities for automotive SMEs and Tier-1 s and also for new players to enter EV supply chains, the global supply chains of major players as well as regional supply chains on the international level. And this technology already has stimulated the appearance of thousands of application developers, especially among SMEs.

According to the 2014 European Automobile Manufacturers' Association's (ACEA's) pocket guide, the EU27 motor vehicle industry employs 3.1 million people in vehicle manufacturing alone. According to inquiries with Chambers of Commerce around Europe, SMEs employ about 55 % of European automotive workforce, some act as Tier-1s, most as Tier-2s, many are providers of specialists to either Tier-1s or directly to OEMs. SMEs participate in all phases of the innovation value chain, including R&D, first concept, vehicle design and modelling, engineering, manufacturing, services and sales. Hence, SME's are an essential part of the European automotive industry and play a significant role in the innovation of the road transport sector.

So far the Internal Combustion Engine (ICE) has been the heart of the automotive industry: the complexity of this industry is such that innovation could be driven only by a few organizations. The advent of electric mobility and the demand of smartness and new forms of integrated transport has changed the context: innovation can be made by many organizations, with SMEs playing new key roles both as suppliers of services and components or directly as vehicle manufactures.

The European Coordination and Support Action "Global Opportunities for Small and Medium Size Companies in Electric Mobility" (GO4SEM) funded by the European Commission in 7th Framework Programme (FP7) aims to give policy advice to sustain and enhance the successful access of European SME's to supply chains in the global markets. The project supports to strengthen the global competitiveness of the European industry being active in the domain of electric mobility by linking the relevant stakeholders, preparing them for and increasing the awareness of the opportunities and challenges of European worldwide developments within electric mobility supply chains.

In this contribution, the electric mobility markets in Europe, USA, Japan, South Korea and China are shortly introduced showing that electric mobility is already conquering markets globally. In the course of the paper, the changes in the EV supply chain compared to the conventional automobile supply chain are briefly discussed in order to concentrate on the role for Tier-1 and SMEs in the international context and to motivate in general the emergence of new opportunities for old and new players. This is followed by a short discussion on the status of the EV supply chains in Europe, USA, Japan, South Korea and China. These informational chapters lead to the discussion of strategies for European SMEs to seize those opportunities. Furthermore, entry points for European SMEs in the specified international markets are elaborated. In the conclusions the main findings are again summarized.

2 The Global Market for Electric Mobility

At the end of 2014, in terms of passenger cars there were about 700.000 EV and plug-in hybrid vehicles (PHEV) on the road worldwide. In this section the market size and development potential of the EU, USA, Japan and South Korea and China will be given. A more detailed analysis of these international markets can be found in the reports on the analysed countries published on the GO4SEM website [1]. It is important to note that small cars are becoming more and more popular [2]. Also the low speed EV market is rather big and forecasted to grow further [3]. As will be discussed later, especially in these markets opportunities arise for SMEs and Tier-1s.

In Europe an accumulated number of about 85,000 BEVs (131,000 including Switzerland and Norway) and 63,000 PHEVs was on the roads end of 2014 [4]. In April 2014, the EU has adopted a directive for the deployment of alternative fuels infrastructure [5] which requires EU Member States to develop national policy frameworks for the market development of alternative fuels and their infrastructure. The EU is also supporting R&D on EVs in multiple coordinated funding programs like the European Green Vehicles Initiative PPP. Furthermore, a number of the member states, as e.g. France, UK, Germany, Spain and many others, follow very dedicated and ambitious agendas to support the development but also implementation of electric mobility. Taking all announcements of EU member states together, the goal for 2020 is set to 813,000 charging stations in the European Union and to have 8834,000 EVs on European roads.

Taking a look at international markets, the US shows the largest market volume with about 285,000 EV and PHEV passenger cars and 9000 PHEV buses sold till 2014 [6]. Its global leading position is due to both, the headway position in related technologies and to the remarkable success of the Tesla Model S introduced on the market in 2012. In order to promote the electric mobility market, the US government has launched several initiatives promoting technological R&D funding, pushing the installation of charging infrastructure and gaining the consumer acceptance and demand for EVs and PHEVs. The US foresee up to 3 million EVs on their roads by 2020 [7]. The US market is currently oriented to mid-sized (1300–1500 kg) and high performing large sized (>1500 kg) vehicles.

The second largest electric mobility market is located in China with an approximate number of 75,000 passenger cars and 35,000 buses sales until 2014 [6]. The automotive industry in China is comparatively young but the most dynamic in the world today. China is pushing even harder on the development of what is called new energy vehicles there in order to gain technological know-how and competitive advance quickly and thus, to occupy a leading position in the international market. China is the world largest producer of batteries but it is still behind Japan and Korea in terms of automotive grade developments. China is focusing on small (<1000 kg) and city fully electric vehicles (1000–1200 kg) while much less emphasis is given to hybrid or range extended solutions. In 2014, also 400,000 low speed electric vehicles like bikes and pedelecs were sold in China [8].

Referring to the latest numbers, 100,000 EVs have been registered in Japan till 2014, rating the market at third place [6]. Japan is one of the leading electric mobility markets, mainly due to pioneering activities in the domain of batteries and the immense success of the Nissan Leaf. The reason for this remarkable success and progress can be attributed to early development and the governmental support of research and development of battery and energy storage technologies. A major share of vehicles on Japanese roads are conventionally powered Kei-Cars (<660 cc combustion engine) [1] a sector in which the fully electric version can be expected to dominate in the forthcoming years.

The industry in Korea has only sold 2000 EVs so far [6]. At present, the Korean EV market together with Japan already holds global leadership in battery technologies, as several of the world's biggest battery manufacturers are domestic. Initially the South Korean government envisioned the ambitious goal to become the fourth largest market for e-mobility in 2015. Due to delay in EV development and production, South Korea decreased their target for 2020 from about 1 million to 200,000 units [9].

3 The Changing EV Supply Chain

Due to the essentially different properties and involved technologies of the electric vehicle as well as the need for new services the set-up of the supply chain for the classical ICE-based vehicle is challenged.

The upcoming electrification of vehicles will result in a shift in the creation of added value in the supply chain. Compared to an ICE based supply chain, where 30 % of the value added is generated through the power train, for a battery electric vehicle the value added counts for 60 % mainly due to the battery [10]. Furthermore, the production of electric drivetrains requires know-how, which is not yet fully built-up on either side, suppliers and OEMs. The progressive integration of new components and systems in the smart EV such as battery cells, power electronics, high-voltage wiring, electric motors and advanced components for comfort, safety and infotainment allow the entrance of new suppliers and even OEMs. Based on a 2012 survey with global automotive executives, KPMG found that electric component suppliers (including batteries) are estimated to be the most important new players in novel automotive supply chains followed by the lightweight high yield stress materials suppliers [11].

A prominent example for a newcomer OEM is Tesla, which started small in 2003 by developing its own technology for electric luxury cars. Tesla is now a world known brand image going to expand its portfolio into other segments. It is worthwhile noticing that Tesla's business is not expected to be profitable before 2020 and that the losses produced so far has been covered by intelligent financial models and the US Government support. Since mass production is a complex process that requires high investments in advance, specific know-how and economy of scale, the current OEMs will continue to play a dominant role in vehicle manufacturing in the future.

New players with potentially high impact are the manufacturers of batteries: the European OEMs lack production capacity and know-how and are hence engaging with (mostly Asian) battery manufacturers, and this rather in a partnership than within a supplier-customer relationship. Thus, decision-making power as well as added value is shifted upward on the supply chain. In addition, the introduction of EVs but also smart (electric) mobility in general requires new services, business models and actors such as Mobility Service Providers or Web. 2.0 brokers. Independently from vehicle category Gartner [12] expects that six percent of the added value from Internet of Things (IoT) technology (defined as benefits to industry from sales and usage of IoT technology) in 2020, ~$114 Bn, will be in the transportation sector. For example there will be a quarter of a billion connected vehicles on the road, enabling new in-vehicle services and automated driving capabilities etc. The EU Cluster Observatory has identified mobility industries, comprising activities that provide products and services which aim to optimise the mobility of goods and people, as a key emerging industrial sector [13].

In addition, it is frequently stated that complete new business eco-systems that include not only the end customer, the automotive supply chain and mobility providers but also utilities, manufacturers of charging stations and authorities, are necessary for a successful development of electric mobility. This may offer innovative new players a chance to participate in this growing market.

4 The State of the EV Supply Chain in Europe and Beyond

4.1 Europe

In Europe there is a traditionally strong automotive industry with OEMs that have an established brand image and the major automotive Tier-1s in France, Germany, UK, Italy, Sweden and Spain. Today, there are electric vehicle models from French, German and Italian OEMs on the market. Automotive suppliers are spread all over Europe. Automotive clusters are often very active in electric mobility and exist e.g. in West-Midlands/UK, Piedmont/Italy, Basque Country/Spain, Stuttgart region/Germany, Belgium, Finland and Romania. The named clusters have been in detail analysed within the GO4SEM project (respective reports will be available on the project website) [1], however there exist many more.

The European automotive supply chain is taking advantage of the open framework of collaborative research and development promoted by the European Commission and it is further strengthened by national programs [14]. This has led to a wide-spread solid know-how in all mobility related sectors and to impressive achievements in technologies for road safety and reduction of air pollution. The European automotive industry is world leader in smart systems like advanced driver assistance systems (ADAS), embedded automotive software such as AUTOSAR and power electronic components. On the other hand the European supply chain misses the leadership in key semiconductor technologies, is weak in the integration of infotainment and has a lack of know-how for energy storage cells and systems production. Existing partnerships on battery and battery related systems with Korean, Japanese and Chinese producers will probably remain in an uncertain state of making and breaking in the next few years.

4.2 USA

Industrial activities in the U.S. relevant for electric mobility are located both in the traditional vehicle manufacturing area in Michigan, mainly by long-established U.S. car manufacturers, and in the high technology industry clusters of the Silicon Valley (sometimes called the "new Detroit"). Particularly new players settle in the West Coast Region, e.g. Tesla Motors in California as a novel site for car industry. In both cases, high integration into the global supply chains of automobile manufacturing is combined with local partners for logistics and engineering services. All traditional U.S. car manufacturers are engaged in developing low emission vehicles and regularly introducing new generations of PEVs on the market. By this, they are following various strategies serving various car segments and developing full electric and plug-in hybrid drivetrains. The U.S. market is highly competitive particularly in the IT and electronics sectors. The U.S. is world leading in semiconductor chips manufacturing and computing technologies having companies such as Intel, IBM, Qualcomm headquartered in the main automotive regions. As well, there is large expertise in connectivity, infotainment systems and technologies for their integration into the vehicle on the U.S. market. Nevertheless, the U.S. EV supply chain faces the same difficulties as described in general in the previous chapter. One issue is the battery production. Also the U.S. is rather involved in assembling final battery packs of imported battery cells from non-U.S. suppliers than in manufacturing its own batteries.

4.3 Japan

Toyota, Honda and Nissan dominate the Japanese market and have a combined market share of more than 75 %. The six largest Japanese car manufacturers (Toyota, Nissan, Honda, Suzuki, Mazda and Mitsubishi) are each part of a large Keiretsu network that comprises the full automotive supply chain. The Keiretsu system is extremely rigid and non-keiretsu firms encounter severe barriers to entry, making it difficult for trade or partnerships. Recently, this has begun to change towards a more Tier-oriented system. Several Japanese battery manufacturers like Panasonic, Toshiba, Sanyo etc. have specialized in producing high performance batteries for electric vehicles. In some cases, the cooperation of battery-producers and car manufacturers is so close that they share development efforts. All renowned Japanese automobile manufacturers plan to continue the production of environmentally friendly car models over the next few years. Japanese car manufacturers develop and produce technologies for electrified drive trains and cover the whole portfolio of vehicles, ranging from hybrids and plugin hybrids to pure battery electric vehicles and fuel cell vehicles. Kei-cars and micro EVs are produced mostly by traditional OEMs (Daihatsu, Honda, Mazda, Mitsubishi, Nissan, Subaru, Suzuki). These are tailored to suit the people's needs and integrate in intermodal transport chains. Important in this context are also electric bicycles which are manufactured by Japanese companies such as Panasonic, Honda, Yamaha, Sunstar, Bridgestone or Miyata, also often in co-operation with international partners such as Bosch or Daum.

4.4 South Korea

The big four Korean automobile manufacturers, named Hyundai-Kia, GM Daewoo, Renault-Samsung Motors and SsangYong, have started to enter the EV-market during the course of the past 6 years with at least one model each. They utilize both technologies, battery electric and plug-in-hybrid drive trains. The launched EVs focus on the small-sized segments mainly. As typical for the Korean economy the four big players are heading a chaebol, a conglomerate of the OEM and its suppliers, enforcing economies of scope. The South Korean industry for automobile spare parts and components that revolved around the big OEMs is rapidly growing. Technology competence of vehicle components such as motors, inverters and condensers is still relatively low, but Korean SMEs are very active in improvement of their technologies. At present, the South Korean automotive industry has a considerable advantage regarding the battery industry, having several of the world's biggest battery manufacturers on the domestic market. Samsung SDI and LG Chem alone have a market share of more than 20 % of the global cell production, specifically, on the lithium-ion battery production side. Further, new energy vehicles in the micro sized segment find great approval among Koreans as practical small city cars. New energy vehicles are produced by CT&T United, AD Motors, Leo Motors and Top R&D.

4.5 China

Currently, there are more than 100 whole-vehicle manufacturers and almost 8000 automotive parts manufacturers in China, located primarily in Southern, Eastern, Northeastern, and central China. At present, the biggest OEMs are either state-owned (FAW, Dongfeng, and Changan) or owned by local governments (SAIC, BAIC, Chery, and Guangzhou Automotive Group). The only private OEMs in China are BYD and Geely. Most Chinese car manufacturers have short knowledge in battery technologies and battery components. On the other side, most of the Chinese battery suppliers have likewise little knowledge on automotive components. In general, there are only a few dedicated battery manufacturers for automotive applications on the market in China which are then mainly controlled by large OEMs. In the context of light electric vehicles, China is both the world's biggest producer and consumer of e-bikes, pedelecs, e-scooters and other low speed e-vehicles. Similar to car manufacturers, most Chinese automotive parts manufacturers are wholly owned domestic companies such as ASIMCO, Wanxiang, Hongteo, Fuyao, Dicastal, Wanfeng. Along with a growing international demand, the domestic aftermarket for automotive components is becoming an increasingly important driver of the industry as well. Today, most of the world's biggest Tier-1 suppliers have established presence in China, such as Bosch, Delphi, Denso, Johnson Controls, Hella, Lear, Magna, Visteon, Yazaki, ZF, Arvin Meritor and TRW. Many Chinese component manufacturers are still lagging behind foreign component players concerning technological know-how and readiness. The competition on China's new energy vehicles market is getting more and more fierce. Many leading foreign brands and Chinese brands have introduced all-electric or plug-in-hybrid cars.

5 Strategies to Seize Opportunities

Although the powertrain of the fully electric vehicle is mechanically simpler than that of the conventional car, SMEs alone or even when aggregated in clusters with large suppliers, cannot challenge the consolidated OEMs on large-scale manufacturing of vehicles. The current European environment does not offer the possibility that, for instance, Tesla could find in the U.S. with continuous support from venture capital, the government and the stock market. The complexity of the business, the expensive homologations and the necessary large investments to reach the market are prohibitive for European SMEs even when addressing specific niches or the retrofitting of conventional motorisation into EVs.

European SMEs that have addressed the M1 electric vehicles challenging the traditional OEMs are either bankrupt or in serious financial problems. Examples for this are manifold. Micro-vett (Modena, Italy) an SME specialised on transforming conventional FIAT vehicles, and MIA from France went bankrupt after being challenged by EV models of traditional OEMs entering the market. CECOMP, an SME who has started in Torino the production of the Blue Car (known as the Bollorè car), could remain in business thanks to the support received by a large group like Bollorè.

In comparison with the M1 sector, SMEs can play a key and major role in the light EVs and micro EV sectors, where adaptation capacities, flexibility and efficiency are required and the conventional automotive industry cannot fully compete. Three-wheelers, low speed vehicles and micro EVs are clearly new growing businesses open to SMEs which can conceive, design, develop, prototype and produce safe, ergonomic and energy efficient vehicles meeting most people needs. Because these emerging markets are large they are attractive for the conventional OEMs as well. These have a variety of advantages besides they can afford failures and large losses for several years, however, the large OEMs lack the speed and flexibility required by these sectors. Excluding Renault, most large OEMs are reluctant to enter the low speed sector (L6, L7) because less profitable (the Twizy is an example in-spite of its relatively high price).

SMEs willing to address this identified niche areas need to be oriented towards new business models, emerging sectors and technologies, new funding models and tools, relevant value chains and innovation stakeholders, customer needs understanding, etc. More than 20 companies addressing the development of L7e vehicles are available in Italy only. Several others can be found in France, Poland, Germany, UK and in most EU countries. Most of them do not observe the mentioned requirement to make the business sustainable and are thus facing big economic problems. A positive example is the Italian company Alkè. They focus on L7e special vehicles, avoid the competition with large OEMs, adopt lead acid batteries, low speeds (<45 km/h) and limited ranges. For these categories of vehicles crash tests are not demanded. Production is made mostly by manual operations that do not assure high automotive quality standards. The presence Alkè has acquired in the global market is an indication of how a good strategy is as much necessary as having a high technical innovation capability.

Regardless of the vehicle being produced, safety, high quality standards (reliability), ergonomics, aesthetics, smartness and low production cost are required to be successful. Hence, when addressing the manufacturing of the final products the collaboration amongst SMEs, large consolidated Tier-1 s and research institutes is strongly recommended. Especially the need for short developing time, high safety and quality standards, low investments and low production cost, impose the need for SMEs to build close relations with Tier-1s to use off-the-shelf components (carry over) and to profit from their familiarity with the technical demand of highly competitive markets. However, to contrast the large vertical supply chains established in Asian countries a new level of collaborative manufacturing should be envisaged across Europe: the collaborations amongst complementary clusters sharing visions and risks.

6 Entry Points into International Supply Chains

Besides entering or forming novel local EV supply chains acting on local and global markets or local parts of global supply chains as described above, SMEs may consider to enter the supply chains of non-EU OEMs in foreign or generally the global market(s). Of course, the same general opportunities within the changing supply chain as discussed in Chap. 3 arise and the comparison of the state of the EU and international supply chain that can be drawn from the information given in Chap. 4 lead to the following entry points into the supply chains of U.S., Japan, South Korea and China.

6.1 USA

Technology fields which appear as good entry points for an engagement in the U.S. PEV supply chain are new power electronics (SiC, GaN) or electric motors for electric traction drives as well as the development of technologies for dynamic and contactless charging and wireless power transfer. Well suitable seems also the field of new hardware and software for interoperable charging facilities or automobiles and the development of new business models for home or workplace charging. There is further need for smart devices for the grid and communication networks as well as for integrated smart systems for the improvement of security and battery management systems. The intention of both Google and Apple to enter the world of electric mobility is a new opportunity for European companies supplying automotive certified systems. In fact whatever degree of automated or partially autonomous driving a new (electric) vehicle will have there will be a need of mass produced (low cost) high quality automotive grade technologies and systems developed within the conventional supply chain. The segment for small vehicles and two-wheelers provides also opportunities for entering the U.S. EV supply chains.

The recent opening of the Tesla patents can be understood as a way to "train" potential suppliers in key technologies in this context. Contrary to the Keiretsu supplier networks in Japan, Tesla has no need to confine the sharing of knowledge to very few business partners it trusts, but can extend it to anybody, mostly as it is does not have to fear any competition. Although, the opening of Tesla's patent portfolio at the same time appears as an intelligent commercial announcement and time will have to tell the real value of the patents, opportunities emerge for European companies. So far power train and energy storage systems parts have been produced in-house but need to be outsourced when the company will grow further in the future.

6.2 Japan

Due to their world-class expertise, European SMEs can have synergetic effects with Japanese industry by entering specific business fields. Japanese technologies for both battery related systems and electric architectures are the world most advanced. Opportunities for business co-operations between Japanese and European companies may exist on complementary product developments to define common market strategies for emerging markets. As seen from the EU side the most attractive cooperation with Japanese Tier-1-suppliers is in battery and battery related systems for vehicle to home (V2H) and vehicle to infrastructure (V2I) integration.

Opportunities for co-operations in the fields of key technologies for the smart community play a decisive role too, e.g. hardware and software solutions for systems integration, vehicle automation and advanced driver assistance systems (ADAS).

The Keiretsu system however is extremely tight-knitted and making it therefore very difficult for European SMEs to access the Japanese markets easily, in particular regarding strategic and innovative fields. Nevertheless, European SMEs can find alternatives, such as partnering with European Tier-1 in Japan or business with Japanese OEMs/Tier-1 outside Japan. Clusters such as the Open Source EV Cluster, the Electric Vehicle Association of Asia Pacific (EVAAP) and the International Partnership for Hydrogen and Fuel Cells in the Economy (IPHE) each deals with different aspects of electric mobility together with international partners.

6.3 South Korea

The South Korean EV market is promising for European companies because of the high interests Korean companies have in joining international projects. Comparatively low entrance barriers for foreign companies are given in the sectors for development of ICE-converted electric vehicles and of infrastructure for EV charging or battery-replacing as of V2G services. South Korean chaebols are rather hard to access though, better opportunities are business cooperations with emerging Korean SMEs; the government recently set up strategies for promoting the growth of SMEs and building of a stable middle class in Korea.

6.4 China

Due to the fact that the Chinese government wants to strengthen its domestic automotive industry, it will probably be easier for big multinational players to access the market rather than SMEs. However, when it comes to the design and development of key components like electric/electronic control systems and other core technologies there is still a lot of potential for SMEs to integrate in the Chinese EV value chains, since Chinese technological development and know-how in these areas lag behind. Moreover, the Chinese government is working on improvement of safety issues, which might be an additional entry point for European companies. Further potential lies in the selling of electric mobility services or technologies with immediate returns. It is recommended, though, to keep in mind the specialties and requirements of the Chinese market, especially the strong governmental market regulation. One important condition for the establishment of business cooperation in China is the contract of a 50:50 Joint Venture initiative with a local company. Since China is one of the biggest markets for the automotive industry these days, a vast number of international OEMs and Tier-1 suppliers have already successfully established business with the country. Gaining them as customers could be a good incentive to follow and also to enter the Chinese market. Several factors, such as cheaper labor, proximity to suppliers and customers, access to natural resources, and cultural similarity support this fact.

7 Conclusions

As could be seen, electric mobility offers many possibilities of growth to EU SMEs both in Europe and in the global market. Although the OEMs and Tier-1 suppliers will stay the main actors, the role of SMEs is becoming more important, e.g. in the supply chain of M1 electric vehicles. Opportunities to integrate in these emerging supply chains exist for SMEs on the local and global level. On the other hand, the sector of light electric vehicles and low speed electric vehicles offers the opportunity for SMEs and Tier-1s to make own developments, since it is falling outside the large automotive OEMs focus.

The formation of new supply and value chains in these fast growing sectors can be envisioned by the collaboration between large conventional automotive Tier-1 s and SMEs addressing the whole value chain, from end-users to technology start-up.

The strength of European automotive Tier-1s is central to the growth of an SMEs driven industry of light electric vehicles. The development cost of state-of-the-art power electronic and smart systems integration necessary to produce advanced light or low speed electric vehicles are usually prohibitive to SMEs. On the contrary large automotive Tier-1s, starting from their usual automotive products, need only minor efforts to also address light and low speed electric vehicles.

European suppliers and manufacturers are facing the competition of strong global players as e.g. Asian state-supported large vertically integrated manufacturing supply chains in innovative areas such as clean mobility, pharmaceuticals, portable electronic devices, and software. Especially Asian integrated clean and smart mobility solutions consisting of new electric vehicles and advanced software algorithms (for vehicles and other personal devices) have the potential to further challenge both manufacturing and the rate of growth of the European smart mobility economy. Hence, new forms of European collaborative production and procurement are needed to address this challenge.

European manufacturing can be strengthened by supporting different EU regions to produce similar products and services by sharing customer needs, technologies and solutions while addressing specific regional and local needs. Thus local production for local challenges may then be adapted to serve also global needs.

Acknowledgements The GO4SEM project is funded by the European Commission within FP7 under grant agreement number 609256.

References

1. www.go4sem.eu
2. The global market for compact cars. Panorama 2015, IFP Energie nouvelles
3. China low-speed electric vehicle industry report, 2014–2017. PRNEWSWIRE, Sept 2014
4. STTP roadmap electrification of road transport: european deployment strategy and action plan for electric mobility. Final Draft 2.0 (2015)
5. Directive on the deployment of alternative fuels infrastructure. COM (2014) 94
6. IEA Implementing Agreement Hybrid and Electric Vehicles (2015) Hybrid and electric vehicles. The Electric Drive Delivers, Annual report for the year 2014
7. IEA (2013) Global EV outlook
8. Forbes (2015) China's other electric vehicle industry
9. Business Korea (2014) Slow expansion: popularization of electric vehicles taking time in Korea
10. Pieper M, Ernst C-S, Technology, value added and supply chains of electric vehicles. http://www.enevate.eu
11. KPMG'S Global auto executive survey 2014
12. Gartner says 4.9 billion connected "things" will be in use in 2015. Gartner Press Release (2015)
13. Emerging industries: report on the methodology for their classification and on the most active, significant and relevant new emerging industrial sectors. EU Cluster Observatory, July 2012
14. Strengths and weaknesses on European and MS level in supporting the emerging new FEV value chains and opportunities for collaboration. Report of the CSA Smart EV-VC, VDI/VDE-IT. (www.smartev-vc.eu)

Part V
Safety Challenges for Electric and Automated Vehicles

A Smart Computing Platform for Dependable Battery Management Systems

E. Armengaud, C. Kurtulus, G. Macher, R. Groppo, M. Novaro, G. Hofer and H. Schmidt

Abstract An important acceptance criteria for electric mobility is the capability to efficiently use the energy stored into the battery cells over the vehicle lifetime. The battery management system (BMS) plays a central role by estimating the state of charge (current energy available) and state of health (degradation due to ageing effects) of the cells. Improvement of the estimation quality has a direct impact on the battery and thus vehicle range. It is the target of the INCOBAT project to improve the BMS by means of new electronic components, new control strategies

INnovative COst efficient management system for next generation high voltage BATteries, http://www.incobat-project.eu/. The research leading to these results has received funding from the European Union's Seventh Framework Programme (FP7/2007–2013) under grant agreement no. 608988.

E. Armengaud (✉) · G. Macher
AVL List GmbH, Hans List Platz 1, 8020 Graz, Austria
e-mail: eric.armengaud@avl.com

G. Macher
e-mail: georg.macher@avl.com

C. Kurtulus
AVL Turkey, Akpınar Mahallesi Tuna Caddesi, Ballıca Sokak No: 1, 34885 Sancaktepe Istanbul, Turkey
e-mail: can.kurtulus@avl.com

R. Groppo · M. Novaro
Ideas & Motion SRL, Via Trento e Trieste 14, Cavallermaggiore, CN, Italy
e-mail: riccardo.groppo@ideasandmotion.com

M. Novaro
e-mail: marco.novaro@ideasandmotion.com

G. Hofer
Infineon Technologies Austria AG, Babenbergerstraße 10, 8020 Graz, Austria
e-mail: guenter.hofer@infineon.com

H. Schmidt
Infineon Technologies AG, Am Campeon 1-12, 85579 Neubiberg, Germany
e-mail: holger.schmidt2@infineon.com

T. Schulze et al. (eds.), *Advanced Microsystems for Automotive Applications 2015*, Lecture Notes in Mobility, DOI 10.1007/978-3-319-20855-8_19

and new development methods in order to achieve cost reduction and performance (driving range) increase. In this paper, the INCOBAT project is presented and important results with respect to computing platforms, control strategy and dependability analysis are discussed.

Keywords Electric vehicles · High-voltage battery · Battery management systems · Multicore computing platforms · ISO26262

1 Introduction

Electrification of the powertrain is a key challenge to reduce pollutant emissions and improve efficiency. However, the user acceptance of electric vehicle is confronted to different challenges such as costs of the vehicle, driving range, or infrastructure support. Several of these challenges are directly connected to the battery, the central element of the full electric vehicle (FEV). The aim of INCOBAT [1] is to provide innovative and cost efficient battery management systems for next generation HV-batteries. To that end, INCOBAT will propose a platform concept in order to achieve cost reduction, reduced complexity, increased reliability as well as flexibility and higher energy efficiency.

First contribution of this paper is to present the INCOBAT project based on its four main research axes (Customer needs and integration aspects, consistent concept and specification, E/E control systems, and improving system maturity). Furthermore, three selected project achievements are presented. The first aspect is the increase of computing resources by the introduction of the INCOBAT BMS computing platform, which is based on Infineon's multicore AURIX CPU. This platform is the required enabler for the design and deployment of more complex control strategies. The second aspect relates to the migration of electro-impedance spectroscopy approach to the embedded control unit, thus providing more insight of the current status of the cells during field operation and therefore more accurate estimation of the battery state. Finally, an approach for security aware hazard analysis and risk assessment is introduced to enable the consistent identification of dependability targets. The different aspects are all contributing to provide a smart computing platform for battery management systems.

The content of this contribution is organized as follow: Sect. 2 provides an overview of the INCOBAT project, while in Sect. 3 the E/E control platform is presented. Focus of Sect. 4 is to discuss the resulting improvements on the control strategy, combining a new electro-chemical impedance spectroscopy and a model-based battery estimation algorithm optimized for multicore platform. After that, the approach for security-aware safety analysis is presented in Sect. 5. Finally, Sect. 6 concludes this work.

2 The INCOBAT Project—An Overview

The aim of INCOBAT (INnovative COst efficient management system for next generation high voltage BATteries, started in October 2013) is to provide innovative and cost efficient battery management systems for next generation HV-batteries. To that end, INCOBAT will propose a platform concept in order to achieve cost reduction, reduced complexity, increased reliability as well as flexibility and higher energy efficiency.

The targeted outcomes of the project are

- Very tight control of the cell function leading to an increase of the driving range
- Radical cost reduction of battery management system
- Development of modular concepts for system architecture and partitioning, safety, security, reliability as well as verification and validation, thus enabling efficient integration into different vehicle platforms.

  To achieve these ambitious targets, the technical approach chosen in INCOBAT primarily relies on the following 12 technical innovations (TI) regrouped into four innovation groups (see Fig. 1):
- *Customer needs and integration aspects*: ensures a correct identification of customer needs and enables an efficient integration into different platforms. This is supported by the use of mission profiles (TI-01)—in order to take into account the different driving styles of the customers, the different traffic conditions in the same scenarios and the different tracks—and by the integration into a demonstrator vehicle (TI-12)
- *Transversal innovation: consistent concept and specification*. This second group targets the optimization of the system architecture and its consistent description over the technologies and over the system hierarchies. This aspect aims at providing a consolidated basis in order to simplify later industrialization of the proposed technologies. This includes the TI-02 "Model-based systems engineering" to improve correctness/completeness/consistency of system specification, the TI-03 "System architecture—efficient partitioning of the functionalities" for system optimization at BMS or even vehicle level and the TI-04 "Integration of multiple functionalities" to reduce the number of electronic control units (and thus related costs) in the vehicle.
- *Technology innovation: E/E control system*: This third group aims at improving the components of the E/E control system. Regarding the electronic parts, it consists of TI-05 "Multicore computing platform for additional computing resources" and the TI-06 "Smart and integrated module management unit". From the software part, this is achieved by the TI-07 "Modular SW platform" and by TI-08 "Improved BMS control algorithms"
- *Transversal innovation: improving system maturity*: This last group targets the evidences related to the trust on the technical solutions with respect to correct operation (TI-10 Design and validation plan including reliability consideration), functional safety and security (TI-09 Definition and integration of safety and

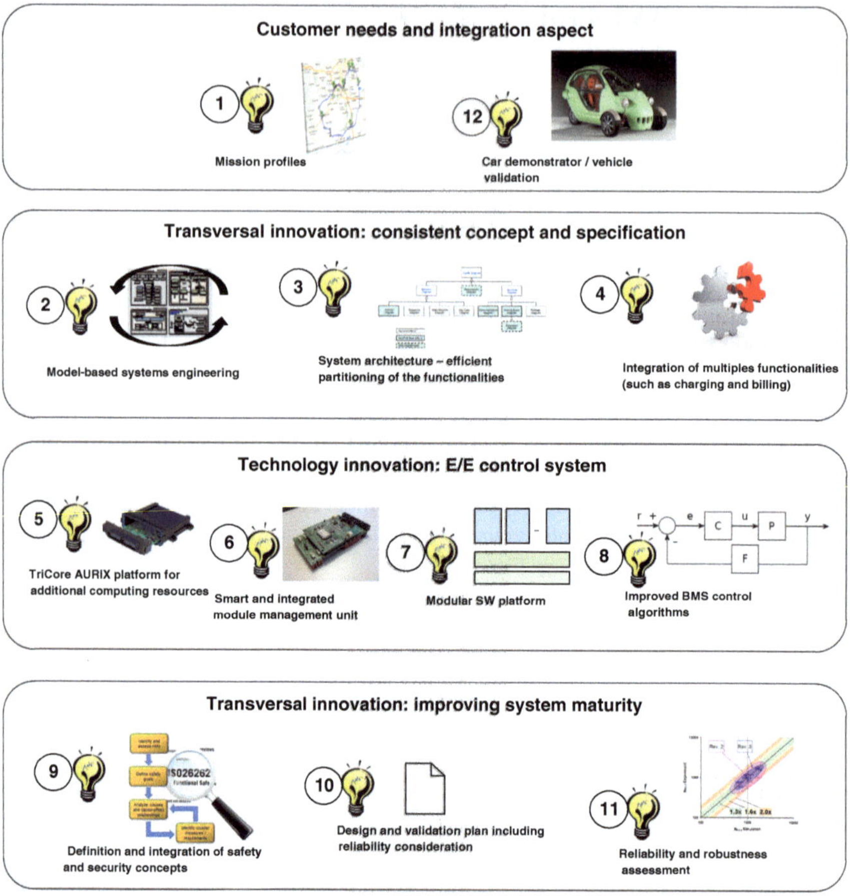

Fig. 1 Technical innovations within INCOBAT

security concept) as well as reliability (TI-11 Reliability and robustness validation). This group of technical innovations is an indicator for the maturity of the proposed technology and further provides information on the efforts required for proper integration and validation of the system.

3 The Smart E/E Control Platform

The goal within INCOBAT is to realize a 'close to production' BMS demonstrator platform to enable the various partners to perform research into BMS architectures and electro-chemical impedance spectroscopy. The benefit of using the latest automotive components in a research platform is that the time taken for the step from research to a production ready solution can be shortened, as techniques employed during research can be more directly applied. Added to this is the ever greater affordable computational power of modern multicore microcontrollers such as the new AURIX TC275T from Infineon. The integration of many BMS functions, with many different requirements into a single CCU is now possible as the parallel processing capabilities of the microcontroller mean that different project partners can supply their software into one common platform and run in an encapsulated processing environment, with freedom from interference' from and to other processes. The objective in INCOBAT is to provide the lowest cost system solution to a BMS by fully integrating all BMS functions into a single ECU.

The CCU is novel in many ways due to the high integration and high functional density. It also supports functional safety up to ASILD so supports even the highest degree of rigor required for mission critical processing. The design integrates the High Voltage (HV) monitoring circuits using an area of the PCB which is galvanically isolated from the rest, again saving cost, reducing component and overhead and increasing performance and reliability. There is also the need for digital signal generation and analogue voltage measurement in the HV domain, so standalone devices are integrated and galvanically isolated by digital interfaces over SPI (Fig. 2).

The INCOBAT BMS CCU is based on the Infineon multicore processor AURIX TC275 based on an innovative multicore architecture [2]. This device supports the concurrent execution of mixed ASIL functions up to ASIL-D [3]. It offers a rich set of peripherals such as A/D converters for data capturing and it has a reasonable

Fig. 2 INCOBAT BMS CCU prototype hardware

number of IOs to support BMS applications. In conjunction with the specific power supply ASIC TLF35584 it is possible to supply the CCU and support ISO26262 requirements with a minimum number of components.

Low and high side drives are available to control contactors for various components such as DC/DC charger. Several digital inputs and low voltage ADC inputs are available. Well known state of the art communication interfaces like CAN-FD, USB and 100BaseT Ethernet are available. For early technology adoption and exploration a novel BroadR-Reach (BrdR) transceiver is provided. An SD card slot is available to record historical data and to support (software) development.

Regarding the SW developments in INCOBAT, a modular development platform is required. Hence, the control strategy and the application software in general is expected to come from different providers and to require different levels of criticality. The activities of SW architect—to define the SW blocks as well as their interfaces—and the activities of SW integrator—integrating the different SW modules and ensuring correct operation of the entire control system—are especially challenging in the context of automotive supply chain with constraints related to functional safety (ISO 26262). A modular platform is required to enable the distributed development and flexible deployment of different control strategies and applications in an efficient way.

For the SW developments in INCOBAT, the proposed common, modular software development platform consists of:

- a layered SW architecture, consisting of several layers and components as well as their interfaces, providing access for the applications to the underlying HW capabilities,
- a suitable SW development tool chain, which supports the application SW developers by means of an effective and consistent development process to seamlessly integrate their particular applications to an overall BMS.

4 Improved Control Strategy Relying on Multicore Computing Platform

Within the project, two algorithms are being investigated in parallel: a model-based estimation method and another completely new one based on electro-chemical impedance spectroscopy (EIS) methodology. From the point of view of the model-based battery state estimation, the main focus is on the analysis of the existing algorithms (currently running on single core computing platform) in order to identify possible improvements with respect to SoX estimation accuracy while making use of the additional computing power of a multicore processor. The main target is to improve the accuracy of the estimations from a group of cells (e.g. a module) down to single cell level. The sensor platform captures or measures certain parameters of the cells. In our case, the sensor platform measures on cell level. The challenging factor here is to run a dedicated instance of the existing model estimation algorithm for each single cell instead of one instance for a group of cells. The required computing power (number of algorithm instances running in parallel) is therefore directly dependent on the modelling accuracy (reduction of the number of cells taken into consideration for one algorithm instance). A higher modelling accuracy provides more accurate information on the status of the cells, thus moving the limits for use of each single cell (and therefore of the entire battery at the end) from a conservative boundary to a more real limit.

Kalman filtering is the state of the art algorithm with regards to accuracy and robustness for SoC-estimation. Because of the fact that the battery is a highly nonlinear system, a modified Kalman filter has to be implemented (nonlinear filter). Detailed information about Kalman filtering for SoC estimation can be found in [4]. The main reason of using a KF is that it considers process and measurement noises to estimate the states, which makes the resulting estimation more accurate when the assumptions of the filter are met. Computational requirements of the algorithm mean that only lumped-parameter estimation is possible on state of the art automotive computing platforms. The high computational power of the multicore platform allows estimating SoC and resistance on cell-level, resulting in a more accurate SoC, SoH and SoF prediction. Furthermore, it can be assumed that the range of the vehicle and life-cycle is more common of the battery can be increased due to the precise estimation approach.

The second, complementary approach followed in INCOBAT relates to Electrochemical Impedance Spectroscopy (EIS). This methodology measures dielectric properties of a medium as a function of frequency. In particular, as applied to the battery cells, the goal of the EIS is to determine the impedance parameters of the cells, determine the state of health (SoH) of the cells as a function of the impedance and determine the parameters of the cell model. In order to successfully determine the EIS spectrum it is necessary to take into account certain inherent problems in the method and the component under test. The EIS analysis is based on the following prerequisites: (a) the system must be linear, (b) the system parameters should not vary over time, and (c) the system is single input, single output (SISO). The lithium battery is not generally satisfying these requirements: therefore, additional assumptions have to be made.

First, the characteristic of a battery is not linear; to calculate the impedance it is therefore necessary to proceed to a linearization; the used technique is to identify a working point on the characteristic and to generate a small perturbation of the electrical characteristics. Second, the battery parameters are not constant; in general, even with open battery terminals (i.e. zero current), the battery voltage varies over time depending on the previous history; to allow the stabilization of the battery voltage it is necessary to wait for the conditions of electrochemical balance in the battery. The required time depends on the temperature (ion mobility) and is estimated in the order of few hours. A measurement made before reaching the equilibrium condition produces data with alterations especially in the lower part of the spectrum; these alterations are more or less evident in function of the imbalance inside the cell. Lastly, in a battery, the voltage does not depend exclusively on the current through it, but also on other parameters, in particular temperature and charge state. During each EIS measurement, these other parameters must remain constant, in order not to influence the output voltage. The battery will then be kept at constant temperature and the current absorbed or supplied should not be such to produce a perceptible change in the state of the cell. In general, it must be ensured that the battery open circuit voltage does not vary within the range of the test, or this change will be computed in the spectrum of impedance.

Regarding stimulus generation, the used approach is based on heterodyne and coherent demodulation: each input signal (the excitation current and the resulting voltage) is multiplied by a sine wave and a cosine wave at a certain frequency, and the convolution is calculated. As a result, the amplitude and phase (at that frequency) of the voltage and current signals can be calculated, and from their ratio the battery cell impedance may be calculated. The same calculation shall be performed in parallel on all the frequency points at which the impedance curve shall be estimated.

This approach reduces the stimulus duration in comparison to chirp signal (a sine wave with variable frequency, sweeping linearly the required frequency range), therefore saving computing time and resources. As a result an HW/SW architecture supporting the EIS algorithm has been defined (see Fig. 3), and a specific EIS daughterboard, providing the correct signal conditioning needed for the EIS algorithm, has been designed.

The effectiveness of EIS approach on BMS platform relies on different attributes such as calculation performances, real time capabilities, accuracy of the sensing circuit, signal generation, and the stability of the battery cell during the measurements, in terms of temperature and chemical reactions. In the course of the

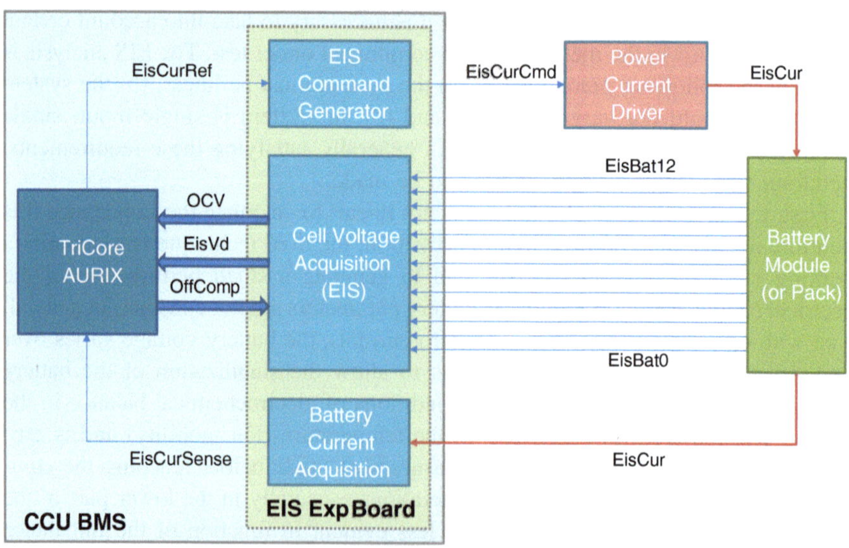

Fig. 3 EIS block diagram

simulation performed so far, we were able to get good confidence that the aspects related to performance and real time capability will not be a problem during the integration into the BMS platform.

The research in the INCOBAT project is currently investigating the relation between the impedance spectra and the SoH of the battery; in particular, experimental work will be conducted on real cells to measure the impedance spectra of the battery when the batteries are in different states. Then, the commonalities of the impedance spectra could be concluded and used to predict the SoH of the battery, and to provide better parameters for the run-time algorithms defining the SoC estimation. Hence, a more accurate representation of cell ageing status (instead of conservative estimations) will be available for SoC estimation, therefore enabling better use of the energy available in the cells. This will have a positive impact on the battery lifetime—especially towards end of life.

The two proposed algorithms well illustrate the INCOBAT strategy: by using improved computing platforms, more powerful algorithms can be deployed, thus leading to more accurate control, further to better use of the energy stored in the cells, and finally to improved vehicle range.

5 Security-Aware Safety Hazard Analysis and Risk Assessment

In the last decade, strong improvements have occurred in the domain of communication and interconnectivity. As a consequence, more and more information from the environment (e.g., road profile, traffic situation) is available and can be used as enabler for major technology steps such as automated driving, active safety or predictive energy management (e.g., based on eHorizon). At the same time, the number, complexity and scope (safety-criticality) of functions implemented in the different control units is rising fast. A strong pre-requisite is therefore to ensure that the delivered service can justifiably be trusted. In the context of BMS, especially the safety and security aspects (more especially security aspects having an impact on safety) need to be considered in a consistent manner in order to enable future technology steps such as predictive energy management (relying on external, non-trusted information) or billing (different stakeholders building their business case on trusted information regarding current battery state).

Safety and Security appear to be two contradicting overall system features, which challenge researchers for decades. Traditionally, these two features have been treated separately, but due to increasing awareness of mutual impacts, cross domain knowledge and fine grasp of commonalities becomes more important. Due to increasing interlacing of systems (such as Car2x in the automotive domain) it is no longer acceptable to assume safety systems immune from security risks. Future automotive systems require appropriate systematic approaches to support security aware safety development. Therefore, a combined approach of the automotive

HARA (hazard analysis and risk assessment) with the security domain STRIDE approach to trace impacts of security issues on safety concepts on system level is presented.

Currently the automotive domain is mainly focusing on the safety-criticality of automotive embedded systems and therefore has several mature methods and processes in place (e.g., hazard analysis and risk assessment (HARA), fault tree analysis (FTA), or failure mode and effects analysis (FMEA)). Also a domain standard for road vehicles functional safety covering the whole product lifecycle (ISO26262 [5]) has been set. On the contrary, several approaches to exposure security design flaws exist in other domains but are not yet propagated in the automotive domain.

The STRIDE threat model approach [6] is an acronym for six security threat categories and uses a technique called threat modeling to review system designs in a methodical way for which type of threat the system is prone to be successfully attacked. First category, *spoofing threats* aim to successfully masquerading as another person or program to gain illegitimate advantages. Second, *tampering attacks* involve malicious modification of data or data orders. *Repudiation threats*, the third category, aim for system lacking in the ability to trace prohibited operations and to counter illegal operations. Fourth, *information disclosure threats* involve the exposure of information, which are not supposed to be accessed. *Denial of service attacks (D.o.S)* simply deny valid services; threats like babbling idiot belong to this fifth category. Finally, *elevation of privilege threats* aim to access, compromise or destroy data that should be available only for privileged user or programs. All these threat classes might have a safety impact (leading to new hazards) when applied to safety-critical applications.

A key outcome of HARA approach is defining automotive safety integrity level (ASILs). The assigned ASIL determines the criticality of System under Development (SuD) defines requirements and measures to be applied for the rest of the systems lifecycle. For the purpose of determining the SuDs ASIL, possible hazards have to be identified which have the potential to put the system in a hazardous state. Afterwards, these hazards are quantified according their potential harm severity (S), probability of exposure (E), and the controllability of the resulting hazardous event (C). The final step formulates high level safety requirements known as safety goals (for more detailed information see [5] Part 3 Annex B).

Threat modeling using STRIDE [6] can be seen as the security pendant to HARA. Key concept of this threat modeling approach is the analysis of each system component for susceptibility of threats and mitigation of all threats to each component to argue that a system is secure. First step of the proposed approach to combining security and safety analysis is to quantify the STRIDE security threats of the SuD in analog manner as done for safety hazards in the HARA approach. Threats are quantified according to the resources (R) and know-how (K) required to exert the threat and the threats criticality (T), see [7].

These three factors determine the resulting security level (SecL). The SecL determination is based on the ASIL determination approach. The quantification of required know-how and tools instead of any likelihood estimation (e.g. of the attacks success or fail) was chosen due to the fact that classification of these factors is more common in the automotive domain and will remain the same over the whole life-time of the SuD. Besides this, the quantification of these two factors is related to the likelihood estimation of an attack to be carried out. The quantification of the threats impact, on the one hand, determines whether the threat is also safety-related (threat level 3) or not (all others). This information is then handed over to the safety analysis methods. On the other hand, this quantification enables the possibility of determining limits of resources spent to prevent the SuD from a specific threat (risk management for security threats). After this quantification these threats may then be adequately reduced or prevented by appropriate design and countermeasures.

In case of safety-related security threats, the threat will be analyzed and resulting hazards evaluated according to their criticality, exposure, and severity. This improves completeness of the required situation analysis of the HARA approach by implying factors of reasonably foreseeable misuse (security treats) in a more structured and consistent way.

6 Conclusion

The enhancement of the battery management system is a key aspect to maximize the use of the energy stored into the cells. However, this development is challenging due to the complexity and heterogeneity of the different technologies involved. An important outcome of the INCOBAT project is the development of a smart, dependable and modular computing platform based on multicore technology. Based on this technology brick, the impact on the control strategy is discussed. Hence, the additional computing performances available enable the refinement of existing control strategies (model-based state estimation) as well as the migration of advanced measurement approaches (electro-chemical impedance spectroscopy) from lab to the control unit. Dealing with safety-critical applications in a connected world leads to the question how to deal with security threats having possible effects on safety. An approach is presented in this work in order to enable the common identification of both security threats and safety hazards, thus laying the basis for common and consistent safety and security development.

References

1. Armengaud E, Macher G, Kurtulus C, Groppo R, Haase S, Hofer G, Lanciotti C, Otto A, Schmidt H, Stankiewicz S (2015) Improving HV battery efficiency by smart control systems. In: Smart Systems Integrations 2015
2. Brewerton S, Grosshauser R (2009) Practical use of AutoSAR in safety critical automotive systems. SAE #2009-01-0748
3. Schneider R, Eberhard D, Brewerton S (2010) Multicore vs. safety. SAE #2010-01-0207

4. Plett GL (2007) Battery management system algorithms for HEV battery state-of-charge and state-of-health estimation. In: Advanced materials and methods for lithium-ion batteries, 2007
5. ISO—International Organization for Standardization (2011) ISO 26262 road vehicles functional safety, part 1–10
6. Microsoft Corporation. The stride threat model, 2005
7. Macher G, Sporer H, Berlach R, Armengaud E, Kreiner C (2015) SAHARA: a security-aware hazard and risk analysis method. In: DATE'15: proceedings of the conference on design, automation and test in Europe, 2015

The Need for Safety and Cyber-Security Co-engineering and Standardization for Highly Automated Automotive Vehicles

Erwin Schoitsch, Christoph Schmittner, Zhendong Ma and Thomas Gruber

Abstract A key long-term trend is towards highly automated vehicles and autonomous driving. This has a huge impact, besides comfort and enabling people not able or allowed to drive, on sustainability of environmental-friendly urban road transport because the number of vehicles and parking space could considerably be reduced if called on command and left behind after use for the next call. This requires a considerable amount of functionality, sensors, actuators and control, situation awareness etc., and the integration into a new type of critical infrastructure based on communication between vehicles and vehicles and infrastructure for regional traffic management. Both, safety and security aspects have to be handled in a coordinated manner, affecting co-engineering, co-certification and standardization.

Keywords Functional safety · Security · Highly automated vehicles · Autonomous vehicles · Safety and security co-engineering · Functional safety standardization · Security-aware safety · Systems-of-systems · V2V communication · V2I communication

E. Schoitsch (✉) · C. Schmittner · Z. Ma · T. Gruber
Digital Safety and Security Department, AIT Austrian Institute of Technology GmbH, Donau-City Straße 1, 1220 Vienna, Austria
e-mail: erwin.schoitsch@ait.ac.at

C. Schmittner
e-mail: christoph.schmittner.fl@ait.ac.at

Z. Ma
e-mail: zhendong.ma@ait.ac.at

T. Gruber
e-mail: thomas.gruber@ait.ac.at

T. Schulze et al. (eds.), *Advanced Microsystems for Automotive Applications 2015*, Lecture Notes in Mobility, DOI 10.1007/978-3-319-20855 8_20

1 Introduction

A key long-term trend is towards highly automated vehicles and autonomous driving which will have huge impact on road transport in the future. Besides comfort and enabling efficient road transport particular in cities even for people not being allowed or able to drive, another fascinating aspect to achieve a sustainable urban transport system is the chance to reduce considerably the number of vehicles required because they could be called on demand and after a drive do not occupy parking space for a long time because they will continue with their next order. This requires not only a considerable amount of functionality, sensors, actuators and control devices, situation awareness etc. but also integration into a new type of critical infrastructure based on communication between vehicles and vehicles and infrastructure, and regional traffic control centers to optimize traffic as a whole and not just locally in the environment of the vehicle. This connectivity results in additional risks because of security breaches being able to impact safety in a critical manner.

The paper will explain how the coordination of safety and security aspects are resolved in different domains and what is proposed e.g. by the Austrian committee to be taken up in the evolving ISO 26262 [1] standard Ed. 2.0., following related activities and considerations in IEC TC65 Ad hoc Group 1 ("Framework for co-ordination of safety and security"), IEC 62443/ISA 99 [2], ETSI TS 102 941:2012 (Intelligent Transport Systems (ITS); Security; Trust and Privacy Management) [3], SAE J3061 (Cybersecurity Guidebook for Cyber-Physical Automotive Systems) [4] and of the Information Technology-Promotion Agency (IPA, Japan: Approaches for Vehicle Information Security) [5].

2 Safety and Security: A "Systems-of-Systems" Challenge

Combined, these road transport systems form connected "systems-of-systems" with new challenges with respect to safety, security, performance and other dependability requirements. Functional Safety Standards for several domains based on the generic basic safety standard ISO/IEC 61508 [6] have evolved since 2000 when IEC 61508 Ed. 1 was completed. The automotive functional safety standard ISO 26262 Ed. 1.0 [1] was completed and published 2011 (parts 1–9) and 2012 (part 10).

The functional safety standards of the first generation did not tackle the challenges of highly connected "systems-of-systems". Particularly the arising security issues were not considered at this time in context of safety. Security in an open vehicle system has become a new factor to be considered in system engineering and safety analysis.

IEC 61508 Ed. 2.0 [6], finished 2010, took as first functional safety standard into account that security may impact safety of a system. Therefore it requires consideration in the risk and hazard analysis phase, with accompanying measures to be undertaken in the following phases; particularly security has then to be reflected in the safety manual.

Although security engineering itself is excluded in the description of the scope of the standard, the standard states that it

> …requires *malevolent and unauthorized actions* to be considered during hazard and risk analysis. The *scope of the analysis includes all relevant safety lifecycle phases.*

The notes definitely address IEC 62443 [2] and ISO/IEC TR 19791 [7] (Part 1, 1.2, k). Security is mentioned in multiple requirements for the safety engineering lifecycle. Security threats need to be considered in the Hazard and Risk analysis:

> The hazards, hazardous events and hazardous situations of the EUC and the EUC control system shall be determined under all reasonably foreseeable circumstances (including fault conditions, reasonably foreseeable misuse and malevolent or unauthorized action). This shall include all relevant human factor issues, and shall give particular attention to abnormal or infrequent modes of operation of the EUC. If the hazard analysis identifies that malevolent or unauthorized action, constituting a security threat, as being reasonably foreseeable, then a security threats analysis should be carried out. (IEC 61508, Part 1, 7.4.2.3).

A security threat analysis should be conducted if a security threat is identified as a potential cause for a hazard. For guidance on security risks analysis IEC 61508 refers to the IEC 62443 series (*Industrial communication networks—Network and system security*) and to ISO/IEC/TR 19791 [7]. It is explicitly noted that malevolent or unauthorized action includes security threats. If security threats have been identified, then a vulnerability analysis should be undertaken in order to specify security requirements (IEC 61508, Part 1, 7.5.2.2).

> Finally, Part 3 requires that all details about security should be included in the safety manual: "The following shall be included in the safety manual: (…) Details of any security measures that may have been implemented against listed threats and vulnerabilities." (Part 3, Annex D 2.4).

Similar concepts are now evolving in IEC 61511, Ed. 2, and ISA TR 840009. Just recently, work on defining harmonized IT security requirements for railway automation was started [8], with the goal to build on the well-known safety certification processes of EN 50129, EN 50159 and integrate security requirements based on IEC 62443 [2].

The initial concept to relate the rigor of security evaluation levels (EALs of Common Criteria) to the potential impact on safety (SIL level) did not find the necessary consensus. Now SLs (Security Levels 1–4) of IEC 62443 [2] seem to be more accepted by industry than the Common Criteria EALs.

In the preparation phase of IEC 61508-3 Ed. 3.0 (Software part), which started Nov. 20–21, 2014, in Frankfurt and was continued in Toulouse March 17/18, 2015, it was decided to look at the ongoing activities in ISO and IEC with respect to "security-aware safety" and to provide more mandatory and informative guidance on a coordinated approach to security in context of functional safety.

In IEC TC65 (Industrial-process measurement, control and automation) considerable concerns arose with respect to the safety impact of security issues in industrial automation systems, since many complex systems of that kind are becoming connected “systems of systems”, particularly by interaction based on wireless connectivity from sensors/actuators to complete plants, grids etc., in maintenance and operations. An Ad hoc Group (AHG1- “Framework towards coordination of safety and security”) was founded to look into the issue and provide recommendations how to handle the co-ordination of security issues in functional safety standards. The kick-off took place Oct. 28/29 2014 at VDE in Frankfurt. In the first meeting overviews were provided by several participants from Europe, Japan, China, US and Australia on ongoing activities and some research projects. E. Schoitsch from AIT provided an overview on several domains and the ARTEMIS projects ARROWHEAD, EMC^2 and SESAMO which had in-depth work provided in the field of security-aware (security-informed) safety. The domains were not restricted to IEC standards areas but included also conceptual ideas from railways (EN 50126/28/29 and EN 50159), Airworthiness standards, Nuclear, Off-shore Platforms and Automotive (including pre-information from safety and security workshops e.g. at Fraunhofer IESE, ISSE WS at SAFECOMP 2014 Florence, ICCVE 2014 Vienna, ISSC Boston 2013 etc.).

The overall question to be discussed and recommendations to be given are: “*How to manage safety and security—in cooperation, integrated, separately? How to certify critical industrial systems taking industrial Cyber-security into account?*”

A short overview on standards’ approaches discussed is provided here:

- Railways (DIN/VDE just updating EN 50129: Pre-standard DIN V 0831-104)—integrative approach (with IEC 62443, SL 1).
- Airworthiness Standards: 3 security standards (DO 326A (E 202A) Airworthiness Security Process Specification; DO 355 (ED 204) Information Security Guidance for Continuing Airworthiness; evolving DO YY3 Airworthiness Security Methods and Considerations)—far reaching separation
- IEC 62859: Nuclear power plants—fundamental principles defined how to include cybersecurity without impacting safety.
- IEC 61511/ISA TR 840009 (draft) proposes the Cyber Security Life Cycle to be integrated with Process Safety Management.
- TC44, Safety of Machinery, electro-technical aspects: separation of safety and security already at requirements level, OEM (integrator) should be the only responsible, not the machinery manufacturer—not appreciated e.g. by ISA or most of the experts.
- Example from off-shore facility: different safety and security levels at different parts of the facility assessed jointly, to be considered in allocation phase.
- IEC 62443 (security levels SL 1–4) versus Common Criteria (ISO 15408, Evaluation Assurance Levels EAL 1–7): IEC 62443 the preferred standard for industrial automation.

The proposal from Austria (AIT) to ISO TC22 SC32 WG 08 presented at the ISO 26262 meeting at VDA in Berlin, Jan. 29/30, 2015, was set up by the authors after the kick-off meeting, taking up ideas as well as some concerns pro and con from AHG1 members.

3 Security and Privacy: An Additional Challenge in Open Safety-Related Systems

3.1 *Security as Challenge in Open Safety-Related (Critical) Systems*

In the past, vehicles were separate units that occasionally interact with other road users in a physical manner only, the responsibility and ability to control the vehicle was totally with the driver (Vienna Convention on Road Traffic, 1968). The "connected car", may it be with other vehicles or infrastructure in its environment, with highly automated up to autonomous driving changes considerably the original approach to safety and responsibility and occasionally of liability. Now the system is transformed into a system of systems with wireless connectivity. This makes the system to an open system, vulnerable because of multiple access points from inside and outside the system.

Vehicles rely now on seamless data exchange and information flow. The increased connectivity and interaction gives rise to new hazards. Hazards and their causes, faults and vulnerabilities are no longer restricted within a single vehicle, and no longer under full control of the driver. Due to connectivity, vulnerabilities and faults from a single vehicle can propagate further, leading to hazards affecting multiple vehicles at once. A malicious attacker might exploit vulnerabilities or tamper exchanged data, causing hazards on a multiple vehicle level, comparable to mass collisions because of bad weather and road accidents. The change of the infrastructure from a passive system to an active system makes it susceptible to security threats which lead to additional safety hazards. For example, an attack on a traffic control system [9] might cause vehicle crashes. Since connected vehicles form a connected system of systems, safety and security must be ensured not only on the sub-system but also on combined system level.

Experimental analysis demonstrated that particularly such open vehicle systems have increased attack surfaces and potentials, allowing misuse and manipulation of in-car systems. Potential manipulations include controlling vehicle sound system, stopping the engine while the car is still running, or jamming the brakes. The targets of the attacks include Bluetooth connection for user devices, long range connection over cellular networks for telematics, in-car Wi-Fi access points, and maintenance access points. In case situation awareness and V2V and V2I/I2V are becoming an important factor in automated driving and traffic management, the attack surface increases even more. The motivation of the attacker rises considerably because of the wide-spread deployment and public interest, and potentially political impact, as road traffic is an important part of our society and life (somehow a "pr/marketing" effect for the successful attacker). Therefore, from a safety engineering point of view, security breaches are new causes for hazards at the vehicle level. With an open system we cannot regard a vehicle system to be safe unless the security of the system is assessed and assured.

3.2 Privacy and Its Conflict with Safety and Security

Connected and cooperative transportation systems generate and exchange a large amount of data, particularly to receivers not always known to the driver at the moment data are generated and sent, and it is difficult to estimate what could be derived from the "big data" concerning personal issues. While tampering with these data could cause hazards, eavesdropping on the communication and unauthorized access to stored data (or even misinterpreted data used by authorized organizations) could breach participants' privacy. An arising challenge is the conflict between privacy and safety and security [10]. To ensure safety, vehicles need to make the information related to their position and movement available to all, as often as possible. For security reasons, certain identity information needs to be included in the exchanged data, enabling verification and ensuring data integrity. For privacy, anonymous communication is preferred, and position and movement data should be exchanged as little as possible in order to restrict location tracking and profiling.

General contradictions between different dependability and security attributes are elaborated in [11]. Dependability in this context includes other attributes as well, e.g. reliability, availability, maintainability, performance attributes, sometimes even sustainability, resilience etc. Public acceptance of such systems as described here requires trust in the system, including not only static assurance as done by safety assessment and certification but also run-time assurance because particularly security requires regular updates because of new security breaches arising whereas safety would require rather "never touch a certified system". Systems-of-systems on the other hand have to be adaptive because of their nature—which also requires a dynamic approach to safety in certain cases ("run-time certification"). These aspects have been described in [12] as an approach evaluated in the EMC^2 project. It was also presented as a possible recommendation to the IEC 61508-3, Ed. 3.0 preparation team for the "security-aware safety" challenge. As a result of these considerations and work described, an integrated approach is needed to resolve these conflicts.

4 Consideration of Security Concerns in ISO 26262: A Proposal

Based on the considerations and work presented before, Austria (particularly elaborated by the authors of this paper) prepared a short proposal for the ISO 26262 (ISO TC22 SC32 WG08) standardization meeting at VDA in Berlin. This was particularly under the aspect that many other standardization groups in different domains and the top-level group of TC65 have taken up the challenge arising from increasingly connected rather open systems and systems-of-systems. It was argued that automotive is even more affected than e.g. rather closed systems with smaller attack surface and of less opportunity or motivation to attack like railways.

"Connected Car", "Car on the Internet", V2V and V2I (vehicle to infrastructure) communication, highly automated driving and—in the end—autonomous driving, are no longer unrealistic developments, so the basic automotive functional safety standard should not ignore these facts when developing Ed. 2.0, which includes many other issues as well, not only busses, trucks and motor cycles.

David Strickland, Chief Administrator for the National Highway Traffic Safety Administration (NHTSA), stated:

> ...electronics systems are critical to the functioning of modern cars, and are becoming increasingly interconnected, leading to different safety and cyber security risks. (…) With electronic systems assuming safety critical roles in nearly all vehicle controls, we are facing the need to develop general requirements for electronic control systems to ensure their reliability and security.

For a better understanding of our proposal, we look at key elements of ISO 26262. The standard [1] is an adaption of the generic safety standard IEC 61508 [6] for road vehicles. The system development according to ISO 26262 starts with the item definition. An item is a system or a combination of systems that realizes a function at the vehicle level. The next step is the hazard analysis and risk assessment. Results of the hazard analysis are the Safety Goals and the corresponding Automotive Safety Integrity Levels (ASIL). In the functional safety concept the safety goals are assigned to architecture elements and the means are defined (fault detection, fault mitigation, transitioning to a safe state or fault tolerance). Then the system development starts. The Safety Goals in the functional safety concept are refined and the technical safety requirements are specified. In the system design, the technical safety requirements are allocated to hardware and software. Common Criteria (ISO/IEC 15408) [6] is an international standard for information technology security evaluation. It defines the process for the specification, implementation, and evaluation of security-critical, high-assurance systems. Common security requirements on a class of devices or systems can be specified as a Protection Profile. The known use of Common Criteria in automotive domain is its application to the assessment of security of in-car digital devices, but rather not safety allocated parts. In the industrial automation domain and in functional safety standards like IEC 61508 (and related ones), IEC 62443 is preferred since it addresses the system and not only devices.

IEEE 1609.2 [13] specifies methods to secure messages in IEEE 802.11p wireless communication, an amendment to IEEE 802.11 standards for wireless vehicular communications.

A gap analysis shows that there are clearly several gaps in current standards for addressing safety and security of connected vehicles. The standards are fragmented and incomplete, typically assuming that the "blind spots" are covered by others.

For example, in ISO 26262, security as a risk factor is not included. It is explicitly stated that every other system is presumed to be functioning correctly (ISO 26262, Part 3, 7.4.2.2.2) and only malfunctions in the analyzed item are considered. While "reasonably foreseeable misuse" is mentioned in the risk assessment (ISO 26262, Part 3, 7.4.3.7, similar as in IEC 61508, but IEC 61508 is drawing conclusions with respect to malicious misuse as well), it assumes misuse (e.g. a reckless driver) without malicious intention (e.g. a hacker). While the effect of failures on other items is considered through the interface definition, this stops at the vehicle boundary. This means that ISO 26262 views hazards and causes only on a single vehicle level. The cause for a hazard lies in a malfunction of an item. Hazards caused by malicious interactions between items or manipulations from the outside are not considered. In addition, safety-critical communications with external entities is also not considered. Another main issue is that there is a lack of standards addressing safety and security in a joint way. There are no links between different applicable safety engineering standards and security engineering standards in the automotive domain. While there are security requirements for different implementations of V2X, there is no coherent approach for including security on a system of systems level, and rather not in context of safety.

The automotive security standards approaches mentioned in the introduction, namely

- ETSI TS 102 941:2012 (Intelligent Transport Systems (ITS); Security; Trust and Privacy Management) [3],
- SAE J3061 (Cybersecurity Guidebook for Cyber-Physical Automotive Systems) [4] and of the
- Information Technology-Promotion Agency (IPA, Japan: Approaches for Vehicle Information Security) [5],

are focusing only on partial aspects of the issue: ETSI on communication security, SAE and IPA Cyber-security on security aspects, but not on the interplay between both, safety and security.

The automotive industry faces a few unique challenges for transforming research results into standards and economical productions while integrating safety and security. There is a strong separation between OEMs as system integrators and suppliers. The development is distributed. The elements in ISO 26262 such as "Safety element out of context" and "Development Interface Agreement" support this distribution. Standards must support this distributed structure. Not only security but also other aspects like privacy requirements must be considered as a new factor for safety standards. Private and sensitive information such as locations and vehicle routes needs to be protected. Conflicting requirements for privacy, safety, and security (and maybe other dependability requirements) need to be solved in the design phase.

The proposal presentation in Berlin at the last ISO TC22 SC32 WG08 meeting January 29/30, 2015 included not only the motivation, but also the most important examples from the before mentioned TC65 AHG1 kick-off meeting for the "Framework towards coordination of safety and security".

The proposal left open, of course, the details which approach should be taken, although a more integrated approach was preferred by the proposer (AIT, Austria). The proposal looked like:

- Cyber-security should be included as an risk factor to be considered during hazard and risk analysis
- If necessary appropriate security measures should be implemented, e.g. include recommendations for fitting security standards into ISO 26262 Ed. 2.0
- Include a requirement consolidation phase to resolve potential conflicts and coordinate safety and security requirements
- Validation of safety concept should consider security concept
- Security has to be considered throughout the whole (safety) life cycle—recommendations to be included where appropriate

Several countries (e.g. Japan, Germany, France, Austria) reacted positive and some pointed already out which approach they would prefer (e.g. from IEC 62859, Nuclear, which provides detailed requirements for coordinating safety and cyber-security without affecting the safety requirements ("Safety first")). There were several objections as well, particularly since some time ago in a meeting the issue was decided not to take up in the safety community because of missing knowledge and experience in security. A small subgroup was defined which should look into the issue again.

Here I want to cite an article from the (UK) Safety-critical Systems Club Newsletter, Sept. 2014, Robert Oates and David Banham (Rolls Royce) on "Safety and Security": "*High-integrity systems engineers will need to expand their skills*!".

5 Conclusions

Connected, highly automated or even autonomous vehicles, communicating with the infrastructure as well, have to be *safe and secure*. This requires for reasons of public acceptance and licensing standards which allow certification. They have to consider safety not only in an isolated context but under security threats as well. The system "vehicle" is becoming part of a "system of systems", safety and security need to be integrated at the vehicle level and the system engineering level. This is an enormous challenge and already taken up by other standardization committees for generic as well as domain specific standards, mainly in extending existing, established standards according to these new challenges, sometimes in starting to create complementary standards to cover the gaps. They provide interesting examples for automotive industry.

IEC 61508 Ed 2.0 (2010) has already provided a first approach by integrating security requirements. Security threats are to be considered during hazard analysis in the form of a security threat analysis. If such security threats have been identified, a vulnerability analysis should be undertaken to specific security requirements. Details should be included in the safety manual. In the preparation phase of Ed. 3.0 further requirements and guidelines are planned for a more thorough treatment of safety and security coordination.

In the Avionics domain, the approach was to impose special conditions on type certificates for specific aircrafts. Now Standards are developed for the specification of processes, methods and instructions for continued airworthiness security.

Railways plan to integrate IEC 62443 security requirements into the domain specific safety standards (EN 50129, EN 50159) so that only one certification for both concerns ("security aware/informed" safety) should be required for lower SLs (security levels).

The automotive industry needs to define a standardization approach which combines safety and security engineering. Conflicting requirements for privacy, safety and security (and maybe other dependability attributes) need to be resolved at design time. Security concerns can be integrated in ISO 26262 for a combined safety and security standard, or ISO 26262 can be coupled to one or more corresponding security standards with a definition of a joint process on system level. There are several options which have to be discussed among stakeholders, but the issue has to be taken up seriously.

Acknowledgements This work received funding from the EU (ARTEMIS/ECSEL Joint Undertaking) through the projects ARROWHEAD (grant agreement n° 332987) and EMC^2 (grant agreement n° 621429) and the Austrian Research Promotion Agency (FFG) on behalf of the Austrian Federal Ministry for Transport, Innovation and Technology (BMVIT).

References

1. ISO 26262:2011/12 Road vehicles—functional safety, parts 1–10, ISO Std
2. IEC 62443, Industrial communication networks—network and system security, IEC Std., many parts, closely related to ISA 99 Stds
3. ETSI TS 102 941:2012, Intelligent transport systems (ITS); security; trust and privacy management, ETSI Stds. (technical specification)
4. SAE J3061 cybersecurity guidebook for cyber-physical automotive systems; SAE Stds
5. IPA information technology-promotion agency, approaches for vehicle information security, Japan
6. IEC 61508:2010—Functional safety of electrical/electronic/programmable electronic safety-related systems, parts 1–7, Ed. 2.0, IEC Std
7. ISO/IEC/TR 19791, Information technology—security techniques—security assessment of operational systems
8. Braband J (2014) Towards an IT security framework for railway automation. In: ERTS 2014, Toulouse, 5–7 Feb 2014. http://www.erts2014.org, Program, Friday 7 Feb, Session 7C.3
9. Cerrudo C (2014) Hacking US (and UK, Australia, France, etc.) traffic control systems, in Defcon
10. Parno B, Perrig A (2005) Challenges in securing vehicular networks. In: Workshop on hot topics in networks (HotNets-IV)
11. Avizienis A, Laprie J-C, Randell B, Landwehr C (2004) Basic concepts and taxonomy of dependable and secure computing. In: Institute for systems research, Tech. Rep.
12. Schneider D, Schoitsch E, Armengaud E (2014) Towards trust assurance and certification in cyber-physical systems. In: 33rd international conference on computer safety, reliability and security, SAFECOMP 2014. Springer, LNCS, Florence, vol 8696, pp 180–191

13. IEEE 1609.2 IEEE standard for wireless access in vehicular environments—security services for applications and management messages, IEEE Std
14. Koscher K, Czeskis A, Roesner F, Patel S, Kohno T, Checkoway S, McCoy D, Kantor B, Anderson D, Shacham H, Savage S (2010) Experimental security analysis of a modern automobile. In: Proceedings of the 2010 IEEE symposium on security and privacy, ser. SP '10. IEEE Computer Society, Washington, pp 447–462
15. Checkoway S, McCoy S, Kantor B, Anderson D, Shacham H, Savage S, Koscher K, Czeskis A, Roesner F, Kohno T (2011) Comprehensive experimental analyses of automotive attack surfaces. In: Proceedings of the 20th USENIX conference on security
16. Schmittner C, Ma Z, Smith P (2014) FMVEA for safety and security analysis of intelligent and cooperative vehicles. In: SAFECOMP 2014 workshop proceedings, computer safety, reliability, and security, springer LNCS 8696, pp 282–288
17. ISO 15408, Common criteria, ISO/IEC Std
18. Schmittner C, Ma Z, Gruber T (2014) Standardization challenges or safety and security of connected, automated and intelligent vehicles. In: ICCVE 3rd international conference on connected vehicles and expo, Vienna, 3–7 Nov 2014

Is Software Coded Processing an Answer to the Execution Integrity Challenge of Current and Future Automotive Software-Intensive Applications?

Majdi Ghadhab, Jörg Kaienburg, Martin Süßkraut and Christof Fetzer

Abstract In upcoming automotive systems, the high integration of safety-critical software and the use of high-performance controllers with limited integrity is a demanding challenge. Innovations like driving automation require significantly higher computational power than it is available via special-purpose controllers equipped with safety features. The qualification of high-performance commodity hardware for use in safety-critical systems becomes desirable. To cope with the dilemma of low integrity of such controllers, Software Coded Processing reliably shifts the detection of execution errors into the application software allowing high diagnostic coverage of processing units' failures.

Keywords Software-intensive systems · Automotive controller · Functional safety · Coded processing

---

M. Ghadhab (✉)
BMW AG, Taunusstr. 41, 80807 Munich, Germany
e-mail: majdi.el.ghadhab@bmw.de

J. Kaienburg · M. Süßkraut
SIListra Systems GmbH, Niederwaldstr. 37, 01277 Dresden, Germany
e-mail: joerg.kaienburg@silistra-systems.com

M. Süßkraut
e-mail: martin.suesskraut@silistra-systems.com

C. Fetzer
Department of Computer Science, Institute of Systems Architecture, Technische Universität Dresden, 01067 Dresden, Germany
e-mail: christof.fetzer@tu-dresden.de

T. Schulze et al. (eds.), *Advanced Microsystems for Automotive Applications 2015*, Lecture Notes in Mobility, DOI 10.1007/978-3-319-20855-8_21

1 Introduction

The increasing extent and autonomy of software-driven E/E systems accentuates the necessity to guarantee their correct functioning. ISO 26262 defines functional safety as the "absence of unreasonable risk due to hazards caused by malfunctioning behavior of E/E systems" [1]. Embedded controllers represent the central part of such E/E systems. Strict safety requirements have to be fulfilled during their lifecycle.

In this paper, we first present current trends of automotive applications and their impacts on embedded computing. Next, we highlight the contradicting requirements of high performance versus high safety integrity versus cost efficiency. While reliable systems typically employ established hardware techniques to detect random faults of processing units, we focus on lower-cost and more flexible software implemented fault detection mechanisms to enhance powerful general-purpose controllers. We propose to apply Software Coded Processing (SCP) as it reliably addresses systematic and random, both permanent and transient, hardware failures. The last section explains the working principle of SCP, its underlying assumptions, the failure model, and the way how this technique can be deployed in a day-to-day development environment. Finally, we present experimental results with focus on utilizing Commercial-Off-The-Shelf (COTS) controllers and trade achievable detection rates off against performance implications.

2 Future Automotive Software-Intensive Systems

Future vehicles are seen as a connected and distributed network of complex software systems. Visions of a future intelligent networking of driver, vehicle, and environment, e.g. in the context of BMW ConnectedDrive [2], require new approaches for architectures of control unit platforms and their interfaces within the overall systems [3].

- **Centralized Domain-Controlled Architecture**
  The trend of continuously integrating and networking additional ECUs is getting strongly restricted by communication, power consumption, space, and cabling. It becomes essential to shift in-car-networking complexity into software and to integrate more functions per computing unit [4]. Large Scale Software Integration (LSSI) and domain controllers are introduced in [5]. An LSSI system centralizes several high integrity vehicle software components onto a single ECU. Domain controllers are capable integration platforms and server ECUs which control several bus systems [4]. Relocating the rising software workload to domain-controllers requires more performance and accentuates safety integrity aspects.

- **Driving Automation**
  Advanced driver assistance systems moving towards fully automated driving need data from the core vehicle network and the evolving sensor and communications technologies [6]. The central computing platform executes data intensive functions and complex algorithms for environment perception, maneuver planning, and motion control.

In order to cope with the trends of software-intensive systems, design engineers of embedded computers have to fulfill following requirements with respect to cost-efficiency:

- **High-Performance Computing**
  Research projects for autonomous driving and robotics use modern personal computers to its full capacity to process the required software-intensive functions. This implies memory consumption in the range of Gigabytes, CPU consumption in the range of GFLOPS, and utilization of hardware acceleration [7].
- **Dependable Computing**
  A platform shall provide computing capabilities with guaranteed timing, reliability, and integrity. Control functions have real-time constraints, must satisfy functional safety requirements such as handling of hardware and software faults in a fail-safe and increasingly fail-operational manner.
  Other design criteria, e.g. scalability, are also gaining high importance. They are not discussed in this paper as we focus on the challenge of performance and dependability.

3 The Challenge of High-Performance, Dependable, and Cost-Efficient Computing

Intended computing platforms require both high performance and high integrity. Available processors on the market have either high performance or high integrity. References [8, 9] provide examples of recent automotive high-integrity controllers. Communication with various hardware vendors proved the lack of high-integrity processing hardware providing the computing performance as required by future software-intensive applications [7].

Typically, architecture trends from desktop, laptop and server computing migrate into embedded microcontroller applications [10]. High volume standard processors have significantly better cost per performance than special-purpose ones. Table 1 lists price and performance values of popular CPUs and SoCs. Freescale's MPC5643L [8] microcontroller, designed for automotive safety-critical ECUs, is taken as a reference. Performance values are quotes from the respective technical specification documents.

The values as of Table 1 indicate that the performance per price can be a magnitude higher for commodity hardware than for automotive-specialized ones.

Table 1 Performance per price for selected CPUs [11]

Vendor	Model	CPU	Price/$	DMIPS	DMIPS/$
Freescale	MPC5643L	Power PC Lock-step	15	250	16.6
Freescale	MCIMX6U5 DVM10AB	ARM Cortex-A9	26	2500	96.2
Texas Instruments	66AK2H12	ARM Cortex-A15	250	19,660	78.4
Intel	Atom N270	n/a	32	3846	120.2
Intel	i7 4770 k	n/a	339	124,850	368.3
AMD	FX-8350	n/a	180	97,125	539.6

Other parameters like power consumption, failure modes and distribution, product availability, and operational limitations play also an important role when evaluating CPUs regarding their suitability for automotive application. Nevertheless, the performance per price motivates the investigation of such processors in order to provide economically priced computing power. A significant difficulty is the circumstance that these controllers are not self-checking and have limited fault detection capabilities. The functional safety of embedded controller includes the safety integrity of the application software and the computing platform. The integrity of the computing platform is typically ensured through self-checks and hardware redundancy. We use the term "execution integrity" to refer to the detection and handling of systematic, permanent, and transient hardware failures and interference leading to safety goal violations of an executed application.

ISO 26262 lists in Annex D of Part 5 safety measures and mechanisms with high diagnostic coverages achievable for processing units. These measures include hardware and/or software implemented fault detection.

The most straightforward way to duplicate and compare a microprocessor is the technique of a lock-step [12]. Each processor is expected to produce the same outputs given the same inputs. Unfortunately, lock-step microprocessors double the computing cycle budget while providing just the same performance as single processing. And they are susceptible to non-determinisms. A number of mechanisms in current CPUs increase non-determinism and might disconnect lock-step CPUs of recent developments [13].

Additional to the potential limitations of hardware-implemented fault detection, relying on general-purpose processors requires safety measures without changes to the hardware architecture. Therefore, we propose to investigate software-implemented and hardware-independent fault detection for future automotive safety-critical systems. To achieve high diagnostic coverage values by software-implemented fault detection, we consider Software Diversified Redundancy, recommended by ISO 26262, and introduce an alternative approach based on Software Coded Processing.

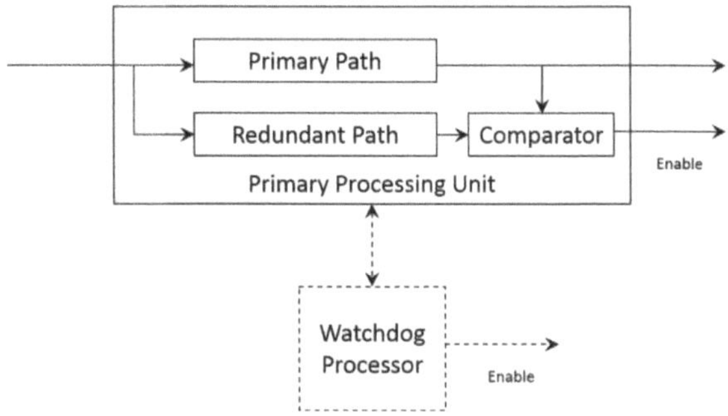

Fig. 1 Software Diversified Redundancy—architecture

- **Software Diversified Redundancy**
  The design consists of two redundant and diverse software implementations in one hardware channel [1]. The redundant path is often implemented using separate algorithm designs and code to provide software diversity. The design must include methods to coordinate these two paths and to resynchronize the paths for transient errors. Due to potential common cause failures, an additional watchdog processor can be used (Fig. 1) and a detailed analysis is required to prove the diagnostic coverage.
- **Software Coded Processing**
  SCP extends the functional code by the ability to detect whether it is correctly executed by its underlying hardware or not. SCP detects transient, permanent, and systematic hardware execution errors with high detection rates. Due to potential common cause failures, the checking of the output validity is performed by an external unit, e.g. a watchdog. SCP permits the efficient execution of non-critical applications and the correct execution of critical applications [14].

Table 2 highlights the main difference between Software Diversified Redundancy and Software Coded Processing. Software Diversified Redundancy requires two (time and effort consuming) implementations, whereas Software Coded Processing requires only one implementation. Considering the achievable diagnostic coverage at lower development cost, SCP deserves further investigation to evaluate its implementation process, detection rates, and performance implications.

4 Software Coded Processing

Software Coded Processing (SCP) is a software technique. The fundamental principle of SCP is the arithmetic encoding of variables, constants, and operations. The result is an end-to-end protection which is hardware independent and continuously present.

Table 2 Software-implemented fault detection techniques

Mechanism	Pro	Con
Software Diversified Redundancy	Allows detection of software faults	Requires diverse software implementations Requires a detailed analysis to prove independence and diagnostic coverage
Software Coded Processing	Detects systematic and random hardware failures Diagnostic coverage can be flexibly configured Detects interference failures between critical and non-critical software	Needs evaluation of performance requirements Requires detailed analysis of dependent and common cause failures

4.1 Execution Errors and Error Model

Execution errors affect the execution integrity of embedded systems' functioning. In this paper, execution integrity is split into hardware execution errors and software execution errors. A hardware execution error is an error that appears in the hardware, e.g. in a processor or in the memory, which potentially affects the execution of the software that is run by this hardware. This could result in a malfunction even if the software is an absolutely correct program (Fig. 2).

Besides hardware execution errors, interference can cause as well erroneous behavior of a given system. This kind of a software execution error results from an unintended interaction between two different programs whereof one could be the operating system, firm- or middleware, or another kind of software. A prominent example of interference is faulty data in memory caused by a faulty write access to a memory region which is supposed to be used only by a given safety-critical application (spatial interference).

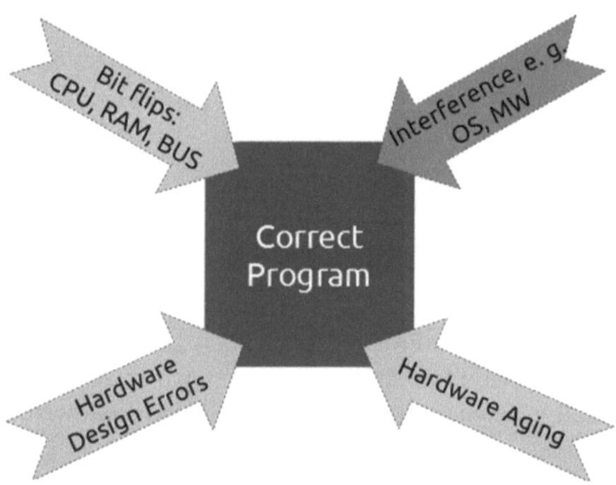

Fig. 2 Execution errors harming a correct program

Execution errors can influence three aspects of an execution: the data flow, the control flow, and the timing. The data flow is the data stored in a system together with all calculations (arithmetic, comparison, etc.) that a safety function performs on this data. The control flow are all decisions that a safety function contains (loops, function calls, etc.). The timing is the timely execution of the safety function. Execution errors can interfere with the timing by making the execution too slow or by stopping the execution completely.

Execution errors are distinguished into transient, permanent, and systematic errors. One prominent example for transient errors are bit flips which could occur in memory, processors, and in bus signals. The characteristic property of transient errors is their randomized occurrence and that they are of very short appearance. A permanent error is an error which appears—maybe only after a distinct period of time—and remains, i.e. it becomes a permanently present error. Hardware aging is one possible reason for permanent errors because it materializes in an irreversible alteration of the electrical behavior of an electronic component. Systematic errors are permanent errors which inherently reside in a given product, e.g. caused by a design error during the development of such a product. Thus, systematic errors appear equally in all products from one development stage.

SCP is capable to detect transient, permanent, systematic errors and errors caused by interference once they propagate into the execution of the software. Errors which do not propagate will not be detected since they have no effect on the execution of the software.

4.2 Arithmetic Encoding

SCP adds information redundancy to a software program to enable it to detect execution errors. The operating principle of SCP is based on a code transformation during which variables, constants, and operations are arithmetically encoded. This code transformation can be realized either manually or via the software development tool SIListra Safety Transformer that works like a compiler and carries out the code transformation in an affordable time.

One part of the generalized procedure of developing an embedded system is writing the functional code. This functional code determines the behavior of the final system. After this functional code is written, it gets compiled and uploaded onto the system. In order to deploy SCP, one additional step is needed: The functional code has to be transformed from its original version into a new version which then contains the arithmetic encoding. As a result, a new program code is generated which still contains the original behavior and, in addition, intrinsically carries the protection via SCP. This new code, functionality-wise identical and SCP-protected, must be compiled instead of the original functional code. As a result, a different binary file is created that has to be uploaded onto the embedded system. Having done that, the system's function remains identical compared to its origin plus it becomes seamlessly and intrinsically protected and safe due to the presence of SCP. Furthermore, safety is achieved independent of the used hardware. By deploying SCP, also COTS hardware can be used in safety-critical applications.

SCP itself is not limited to one kind of arithmetic encoding. There are different encodings which are providing different degrees of protection. The most known encoding is the AN encoding [15]. The basic principle of this encoding is based on the multiplication of any value with a constant A and, to validate the correctness of calculations or results, the check whether a result is still a multiple of this constant A. Values that are not multiples of A are considered as invalid. With SCP, all operations in a program must work with these encoded values. An execution error produces invalid values.

As an example, the numbers 2 and 3 shall be added in the original program code. The expected result is 5. If protected by AN encoding with A = 7, the sum 2 + 3 turns into 14 + 21. Without an execution error, the result is 35 and a valid value because it is a multiple of 7.

4.3 *Diversified Encoding*

SCP can be deployed in different ways during product development. A lean approach is the Diversified Encoding based on the AN encoding [16]. Figure 3 depicts the block diagram of Diversified Encoding.

Diversified Encoding is based on two distinct executions of the same safety function on one channel. These two executions are the native execution and its corresponding encoded execution.

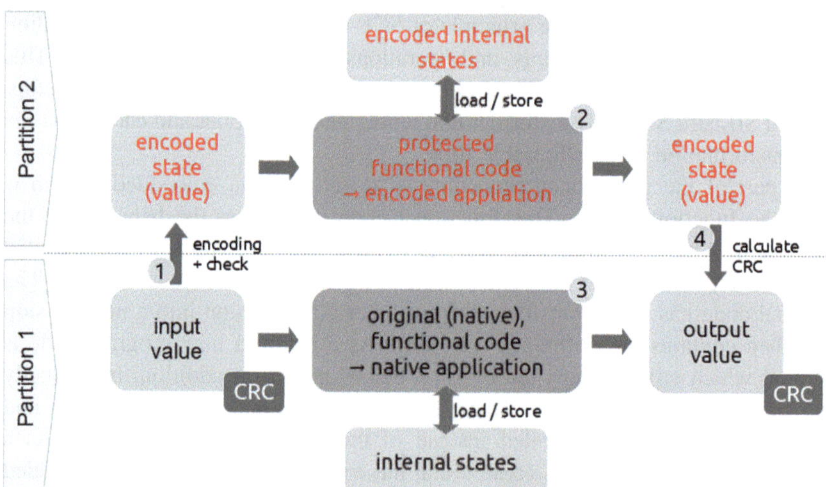

Fig. 3 Diversified encoding—block diagram

- **Native Execution**
  The native execution is the result of the original source code of the safety-critical function. This source code operates on the native input values and native states. The native execution changes only native states. The result of the native execution is the native output.
- **Encoded Execution**
  The encoded execution is based on the encoded variant of the safety-critical function. The encoded execution operates on encoded input values and on encoded states. It produces an encoded output.

Both executions are completely distinct computations but operate on the same values. The encoded input values are the encoded variants of the native input values. The source code of the original, native code is used to generate the encoded source code thereof.

The data flow starts at step (1) with the native input values. The native input values are encoded to become the input values for step (2). In this step, the encoded function is computed. It reads the encoded input values and the encoded internal state. It performs its calculations, updates the encoded internal state, and produces the encoded output values. In step (3), the native function is executed. It reads the native input values and the native internal state. It updates the native internal state and produces the native output values. The checksum of the native output values and the checksum of the encoded output values are computed in step (4). If both checksums are identical, the calculation of the native data flow is considered to be correct. The native output values are taken for the subsequent data processing.

4.4 Experimental Results

Since COTS hardware is experiencing higher attention even for safety-critical automotive applications, respective investigations were carried out. A test software was run on different COTS hardware. The symptoms of execution errors were injected into the test software during run-time, i.e. while the test software was executed on the different COTS hardware. The failure injection was done by the injection tool SIListra Safety Evaluator [17].

Figure 4 visualizes the results from an experiment that was run on a PC with an Intel i7 processor. The test software consisted of three parts: input, an interpolation of characteristics,[1] and output. The interpolation part was defined as safety-critical and, thus, was subject to the failure injection after it was protected via SCP (AN encoding with Diversified Encoding). A grand total of over 300,000,000 failure symptoms were injected and analyzed. The results were categorized into:

---

[1]German: Kennfeldinterpolation.

Fig. 4 Diversified Encoding —experimental results

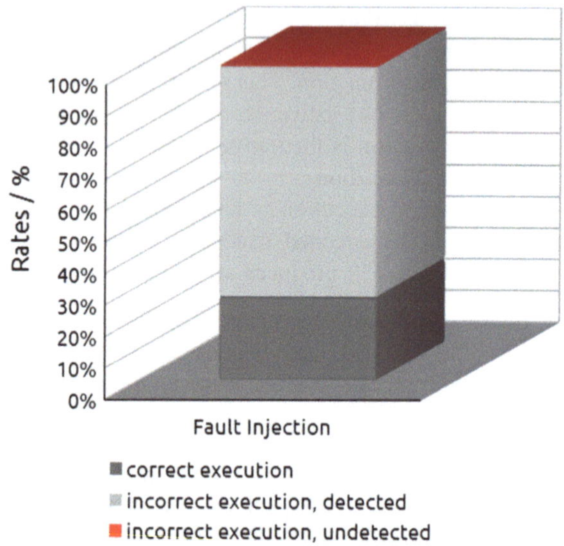

- Correct executions: The injected failures did not falsify the execution and results of the test software.
- Incorrect executions, detected: The injected failures falsified the results and SCP detected the falsification. Cases when the test software was aborted were also counted in this category.
- Incorrect executions, undetected: The injected failures falsified the results but SCP did not detect the falsification.

Only 0.002 % of the injected failure symptoms were not detected. In other words, 99.998 % of the injected errors were either detected or had no impact on the correctness of the executions.

Next, an equivalent experiment was carried out which made use of a new kind of encoding (Fig. 5). Diversified Encoding was not used in this experiment. Instead, only this new kind of encoding was used. As a goal, it was targeted to have less than 1 % of undetected failures while requesting a minimum performance adder to the system. The test software was replaced by Bubble Sort and CRC as test programs for memory-intensive load and CPU-intensive load. Figure 5 shows the results of this experiment. Although not fully optimized yet, this new encoding provides as well rates of "undetected" in the range below 1 %: 0.88 % for Bubble Sort and 0.24 % for CRC (both with control flow check, CFC). The slight reduction in the ratio of undetected execution errors—in absolute figures still on ASIL D levels—results from the circumstance that this new encoding does not use the Diversified Encoding (Fig. 4 vs. Fig. 5).

The evaluation of the performance requirements of this new encoding was carried out on different hardware platforms allowing to derive indications which detection rates could be achieved with which performance requirements.

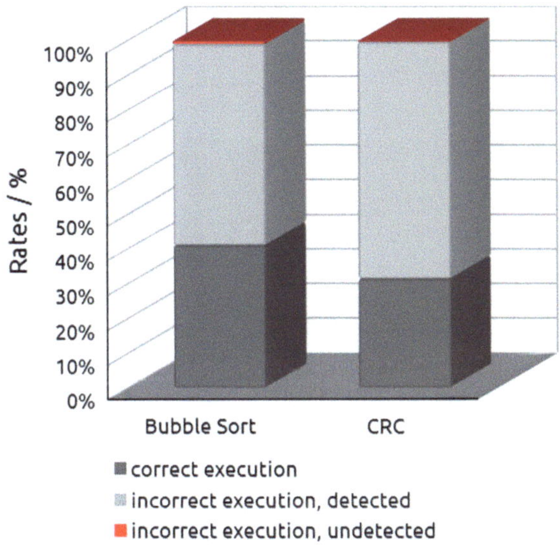

Fig. 5 New encoding—experimental results

Because ARM and INTEL architectures currently gain interest in the trend towards COTS hardware in safety-critical applications, two 32 bit ARM and two 64 bit INTEL architectures were selected. Table 3 tabulates the main characteristic properties of the investigated COTS hardware platforms.

Figure 6 visualizes the performance requirements. All values were normalized to the performance required without SCP. No overall rule-of-thumb number can be derived to quantify the performance requirements. The extent of the required performance depends on different parameters which can differ from case to case. The explicit composition of the native code and the used processor architecture influence significantly the resulting performance requirements. Since 32 bit code transforms into 64 bit code via SCP, 32 bit processors require extra performance for the processing of 64 bit (encoded values). The result that the control flow check (CFC) requires performance while it provides additional protection is known.

Bottom line, SCP improves the safety of systems it has been deployed to. It detects execution errors with high detection rates allowing to be used in ASIL D applications. As a pure software technique, SCP is independent of the underlying

Table 3 Investigated COTS hardware—performance and resources

	ARM v6 k (Raspb. PI)	ARM Cortex-A7	AMD E-450	Core i7 3720QM
Architecture	ARM	ARM	INTEL	INTEL
Bit width	32	32	64	64
CPU clock/GHz	0.7	1.0	1.65	2.6
Memory clock/GHz	Not available	Not available	0.508	0.65

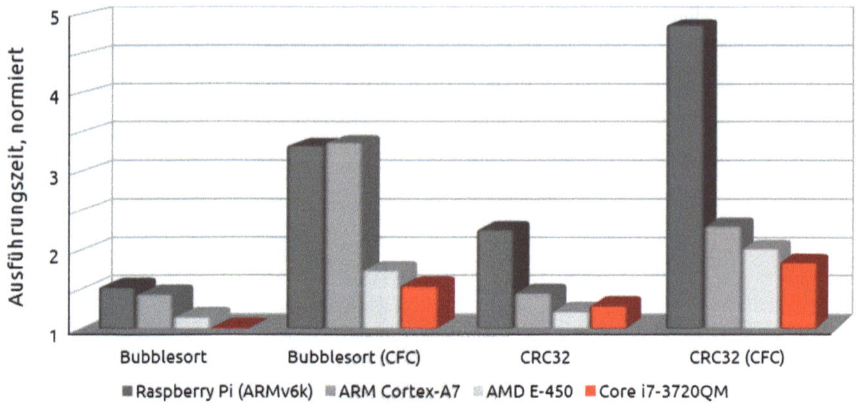

Fig. 6 New encoding—performance requirements

hardware and provides a continuous protection against execution errors. Thus, SCP is an ideal technique for domain-controllers as well as COTS hardware.

On multi-core systems, SCP can be used to implement fail-operational behavior: Two channels are protected with SCP. SCP detects whether a failure occurs in the channels. When one channel fails, operation can continue with the results of the correctly executed channel.

Furthermore, SCP can replace or supplement software mechanisms for safety-critical systems such as instruction set tests, cyclic memory checks, and redundant data storage. SCP provides the same detection capabilities as these software mechanisms in addition to the other advantages already mentioned.

5 Conclusion

SCP enables commodity high-performance processors to be used within safety-critical automotive applications with respect to execution integrity. We expect a tolerable performance impact of SCP when software is accurately split into critical and non-critical and adequate hardware is used. Hence, Software Coded Processing is a suitable and already available solution to the challenge of computing performance and execution integrity for future automotive applications.

References

1. International Organization for Standardization (2011) ISO 26262: Road vehicles—Functional safety, 1st edn. International standard
2. BMW (2014) BMW vernetzt die Freude am Fahren. http://www.bmw.de

3. Michel H-U, Kaule D, Salfer M (2012) Vision einer intelligenten Vernetzung. BMW AG in elektroniknet.de
4. Gut G, Allmann C, Schurius M, Schmidt K (2012) Reduction of electronic control units in electric vehicles using multicore technology. ForTISS GmbH, Munich; Audi Electronics Venture GmbH, Gaimersheim; Springer, Germany (2012)
5. Reinhardt D, Kucera M (2013) Domain controlled architecture, a new approach for large scale software integrated automotive systems. In: PECCS—International Conference on Pervasive and Embedded Computing and Communication Systems
6. Ainhauser C, Bulwahn L, Hildisch A (2013) Autonomous driving needs ROS. BMW Car IT GmbH, ROSCon, Stuttgart, Germany
7. Bulwahn L, Ochs T, Wagner D (2013) Research on an open-source software platform for autonomous driving systems. BMW Car IT GmbH, Munich, Germany
8. Baumeister M (2010) Addressing safety standard requirements for IEC 61508 (SIL 3) and ISO 26262 (ASIL D) with the MPC5643L 32-bit power architecture® microcontroller. Freescale Semiconductor, Inc.
9. Ben Cheikh L, Verma A (2014) Safety joins performance. Infineon Technologies AG
10. Circello J (2011) Rationale for multicore architectures in auto apps. Freescale Technology Forum
11. Fritzsch J (2014) Software-based Controller Integrity in safety-critical automotive systems. Master thesis, BMW Group and Technische Universität Dresden
12. Beckschulze E, Salewski F, Siegbert T, Kowalewski S (2008) Fault handling approaches on dual-core microcontrollers in safety-critical automotive applications. Embedded Software Laboratory, RWTH Aachen University, Germany
13. Bernick D, Bruckert B, Del Vigna P, Garcia D, Jardine R, Klecka J, Smullen J (2005) Nonstop advanced architecture. In: Proceedings of the International Conference on Dependable Systems and Networks (DSN), Hewlett Packard Company
14. Wappler U, Fetzer C (2007) Software encoded processing: building dependable systems with commodity hardware. Technische Universität Dresden, Department of Computer Science, SAFECOMP
15. Schiffel U, Süßkraut M, Fetzer F (2009) An-encoding compiler: building safety-critical systems with commodity hardware. In: Proceedings of the 28th International Conference on Computer Safety, Reliability, and Security. SAFECOMP 2009. Technische Universität Dresden, Department of Computer Science, Springer, Germany
16. Süßkraut M, Kaienburg J, Schmitt A (2015) Safe program execution with diversified encoding. SIListra Systems GmbH, In: Embedded World Conference 2015, Nuremberg, Germany
17. Süßkraut M, Kaienburg J (2015) Safety-critical smart systems with software coded processing. SIListra Systems GmbH, In: Smart Systems Integration 2015, Copenhagen, Denmark

MIX
Papier aus verantwortungsvollen Quellen
Paper from responsible sources
FSC® C105338
FSC
www.fsc.org

If you have any concerns about our products,
you can contact us on
ProductSafety@springernature.com

In case Publisher is established outside the EU,
the EU authorized representative is:
Springer Nature Customer Service Center GmbH
Europaplatz 3, 69115 Heidelberg, Germany

Printed by Libri Plureos GmbH
in Hamburg, Germany